Specification Clauses for Rehabilitation and Conversion Work

Levitt Bernstein Associates and
Anthony Richardson and Partners

The Architectural Press Ltd: London

This book has been prepared in the offices of Levitt Bernstein Associates and Anthony Richardson and Partners.

The illustrations are by Gus Alexander BA. Dip. Arch, RIBA and the cartoons by Ian Layzell.

The contents are based on the experience of the authors, and most of the methods advocated are tried and traditional ones. However, the nature of rehabilitation and conversion work, subject as it is to local variations in the way buildings were originally put together, means that neither the authors nor the publishers can guarantee that their suggestions will work across the board in every instance. They have, rather, to be intelligently applied to the conditions of the job in question.

The authors are grateful to the following people for their advice and/or contributions to the book:

Andrew Garnett, AA Dipl, M. Des (RCA), RIBA
David Graham, BA, B. Arch.
Patrick Hammill
Simon Johns, BSc, Dip Arch
Peter Sanders, BA, B. Arch
Crispin Wright, BA, B. Arch

First published in 1981 by the Architectural Press Ltd, 9 Queen Anne's Gate, London SW1H 9BY

ISBN 0 85139 582 1

Printed in Great Britain by Mackays of Chatham

Table of Contents

Introduction

In recent years rehabilitation has become a more significant proportion of the building industry's work than before. Many experienced architects have found themselves commissioned for rehabilitation jobs for the first time. To the uninitiated the first project can be a terrifying encounter with the unknown for which a lifetime's work on multi-million pound projects will have been no preparation.

This is not surprising, as it has become a specialised field with its own body of knowledge and procedures. The first work should be done with caution and after gathering as much back-up information as possible.

This book, which deals with residential rehabilitation, is an accumulation of some of that knowledge and we hope that it will help to take the terror out of the first few steps into this rough and tumble area of the building industry.

The preparation of building documents for the rehabilitation of residential buildings is a wider subject than it might at first appear. First there are a great variety of existing buildings in their size, plan form, construction and condition and countless alternative ways of converting and repairing them. Second, there is no predominantly used method of describing the work to be done to them. How then, is it possible to write standard or even model specification clauses for the practitioner to use? If those terms mean a comprehensive or encyclopaedic volume, our answer is that it is probably not possible. Even if it were possible, it would be a vast tome, so unwieldy as to be of little practical value.

In compiling this book we have attempted to produce a guide or example rather than an endless list of rarely used specification clauses. We hope it is a practical and useful book but as such it does have its limitations.

It has been compiled and used over a number of years by two firms of architects, Levitt Bernstein Associates and Anthony Richardson and Partners. Both practices were and are, amongst their other work, engaged in the conversion of existing houses, groups of houses and blocks of flats in London for Housing Associations, Local Authorities, developers and private clients. The house types for whose conversion and repair it has been developed, cover a great variety of Victorian buildings as well as some Georgian and Edwardian examples.

The content of the book therefore reflects the necessary repairs of those building types, the coversion requirements of those clients and the prejudices of the authors. Anyone using this book who is dealing mostly with half-timbered thatched buildings or Norman keeps will find the contents less useful than someone dealing with Victorian brick buildings. It is however organised in a flexible way to allow for variation.

It is, quite simply, a 'library of clauses' that can be selected, adapted or rejected for the particular specification that is being compiled. It should be understood that this book is a guide to the Schedule of Works, which is just a part of the total documentation that is normally required to describe the building work.

In the authors' practices the documents that are sent to tender when the job is not of sufficient size to require a Bill of Quantities consist of:

1 Contract Conditions and Preliminaries

2 Materials and Workmanship

3 Schedule of Works

4 Subcontract Specifications (if required)

5 Drawings.

Items 1 and 2 are standard pre-printed documents that, with small insertions, can be used for most jobs. The Materials and Workmanship section is based on the National Building Specification Small Jobs Version, which is of invaluable assistance, particularly when preparing one's first few sets of documents. The Schedule of Works is a detailed description of the particular work to be carried out.

It is meant to be both a pricing and building document. The builder is to set a price against each numbered clause when tendering. Unlike a Bill of Quantities, the clause clearly describes a building operation as a whole so that it can be understood and used on site. Each clause is then broken down into sections sufficient to identify particular parts of the operation by trade and material or with quantities where applicable. It carries equal importance with the drawings as a site document. When the job is of sufficient size to warrant a Bill of Quantities one may still prepare a Schedule of Works with the drawings. It is used by the QS in preparing his bill but is not used as a tender document. However, it is issued to the contractor as a site instruction, as it is a much clearer description of the work than the traditional Bill of Quantities.

The Schedule of Works consists of 11 sections:

1 General Notes

2 Protection/Demolition/Preliminary Investigations

3 Drainage

4 External Works

5 External/Party Walls

6 Roofs

7 Floors

8 Internal Walls

9 Internal Stairs

10 Fixtures and Fittings

11 Plumbing.

In addition there are the following schedules:

1 External Doors

2 Internal Doors

3 Windows

4 Plaster Repairs

5 Internal Decoration

6 External Decoration

7 Floor Finishes

8 Ironmongery

9 Kitchen Units

10 Sanitary Fittings.

It is intended that the Schedule of Works written with the aid of this book will be in the same order as above. Each section draws together building operations on each major area of the fabric, and follows approximately the order in which the contractor will carry out the work.

The procedure for use is to start at page one and carry on to the end selecting only those clauses that are required and, where necessary, adding new clauses. Next the door, window, finishes and fittings schedules are completed. The combination of the two is the complete Schedule of Works, excluding the specialist specifications for central heating, electrical and gas installations. There are more detailed instructions on the following pages.

When the book is used in the authors' offices they have found that approximately 80% of the final Schedule of Works consists of the standard clauses and the remaining 20% is written for the particular project, or adapted to particular circumstances. This, of course, varies from project to project.

A number of clauses have drawings related to them. These are of an explanatory nature and are intended to clarify the clause. The authors have a series of 'Standard Details' which can be adapted and issued in commonly recurring situations.

As they use it, it is not a static document; clauses are revised in the light of experience or new materials and gaps are filled. The authors have not put into the present work every new clause that is written but only those they feel will have repeated use and thus save time and set a standard to be maintained. The book has been written and printed so that those who buy it can put it to similar use. The book can be used as a basic document that can be altered to suit specific needs. It can then evolve into a source that will be tailored to the particular requirements of each individual user.

Instructions for use

The working drawings and details will have been completed and annotated before commencing the specification.

The book is arranged in eleven sections of specification clauses followed by ten schedules. It is intended that the writer start at page one, selecting the clauses required and inserting new clauses where necessary.

Each page has two columns. The left-hand column contains the actual clause to be used with explanatory sketches. The right hand column contains comments about the use of the clause as a whole or specific advice about items within each clause. This column is not intended for reproduction in the final document.

The reader should note that certain British Standard and Code of Practice references have been included in the clauses. They are not comprehensive. It should be noted that these numbers change from time to time and should be added to or revised as necessary. Trade names have, for the most part, not been included. The few that remain are those that are in such common usage that clarity would have been sacrificed by their omission. For daily use of this book the user should amend the clauses to contain the product he wishes to specify. The appendix describes those items which have been abbreviated in the text. They should be included in the materials and workmanship clauses.

The clauses

Each clause contains a title, the word 'location' and one or more subclauses, some of which are shown as options.

When writing a Schedule of Works by hand the simple procedure to follow is to decide whether or not the clause is to be used. If it is to be used, then:

1 Write down the number and title of the clause

2 Write down the word 'location' and then describe the actual location, or locations, within the building

3 Write down the number of each subclause to be included. In many cases all of the subclauses will be required, in others one or more may be unnecessary. In this case just omit the numbers of the subclauses not required

3.1 Some subclauses have 'options'. These are written for example as '4' or 'option 4'. Select which of the two is applicable and just write down either '4' or 'option 4'

3.2 Some subclauses have gaps where words or numbers are to be inserted as noted in the comments columns. In this case write down the subclause number and follow it by the insert you require

4 Decide if an additional subclause is required and just add the next number and write out the new subclause in full

5 Decide whether or not an entire new clause is required. If so write it out using the standard format.

The following page shows **(A)** one clause as printed in the book, **(B)** how the writer is to write it and **(C)** how it appears in the final typed specification.

(A) As printed in the book

7.11.1 Ducted air vent	General
Location:	*Some authorities prefer a separate fresh air supply direct to each bathroom or WC as well as a ventilated lobby and mechanical ventilation. Some authorities ask for a ventilated lobby with a positive airflow to bathroom via grille or similar. Check for particular Environmental Health Requirement.*
1 Brickwork: Cut opening and fit 230 × 150 mm terracotta airbrick on external face of wall. Make good brickwork.	*1 Select either 1 or option 1.*
Option 1 Brickwork: Cut opening, take asbestos pipe to external face of wall, fit wall cowl by .... to pipe, make good brickwork.	*Option 1 Insert name of manufacturer.*
2 Item: Between joists clip 100 mm diam. asbestos pipe grouted to airbrick or cowl. Right angle bend cut to bring end of pipe flush with ceiling.	*2. Select either 2 or option 2.*
Option 2 Item: Between joists clip 100 mm diameter asbestos pipe grouted to airbrick or cowl. Right angle bend and upstand where shown with further right angle bend to bring end flush with external face of partition, or bath panel.	
3 Plaster: Cut ceiling and fit 230 × 150 mm plaster louvre ventilator. Make good plaster.	*3 Select either 3 or option 3.*
Option 3 Item: Mechanical extract fan by electrical subcontractor.	
4 Carpentry: Take up floorboards as necessary, face joist each side of pipe with 9 mm fire-resistant board, cut 9mm short of floor over. Lay 9 mm fire-resistant board over joist space sitting on fire-resistant board linings. Refix floorboards.	*4 Is the appropriate clause when the vent pipe perforates a half hour fire resisting ceiling over.*
5 Item: 225 × 75 mm white plastic air vent to each side of bathroom door, hole cut in door to suit.	

(B) As handwritten

7.11.1 Ducted air vent

Location: Room 6

Option 1 John Doe Cowl Co Ltd

2

Option 3

4

5

(C) As finally typed

Ducted air vent

Location: Room 6

1 Brickwork: Cut opening, take asbestos pipe to external face of wall, fit wall cowl by John Doe Cowl Co. Ltd. to pipe, make good brickwork.

2 Item: Between joists clip 100 mm diam. asbestos pipe grouted to airbrick or cowl. Right angle bend cut to bring end of pipe flush with ceiling.

3 Item: Mechanical extract fan by electrical subcontractor.

4 Carpentry: Take up floorboards as necessary, face joist each side of pipe with 9 mm fire-resistant board, cut 9 mm short of floor over. Lay 9 mm fire-resistant board over joist space sitting on fire-resistant board linings. Refix floorboards.

5 Item: 225 × 75 mm white plastic air vent to each side of bathroom door, hole cut in door to suit.

The Schedules

The Schedules are filled out by hand in accordance with the notes following each schedule at the back of the book. They are then duplicated and bound in with the other documents.

Transcription to final typescript

When the handwritten Schedule of Works (B), has been completed, as described above, it can be transcribed into the final document, (C), by a typist with a copy of this book for reference.

Alternatively the contents of the book can be put on tape and stored in a word processor and the insertion and deletions made using a visual display unit either by a typist with a copy of (B) or by the specification writer direct.

Numbering

The re-numbering of sections, clauses and subclauses should be as follows:

Each section should be numbered as in this book and each clause within each section should be re-numbered as a division of that number. Each subclause should be re-numbered in sequence as they appear in their final form.

Eg. '3. External Works' would be followed by Clauses 3.1, 3.2, 3.3 and so on with subclauses being numbered .1, .2, .3, .4 and so on.

Schedule of Works Clauses

1 General

1.1 General Notes

1.2 Schedule Notes

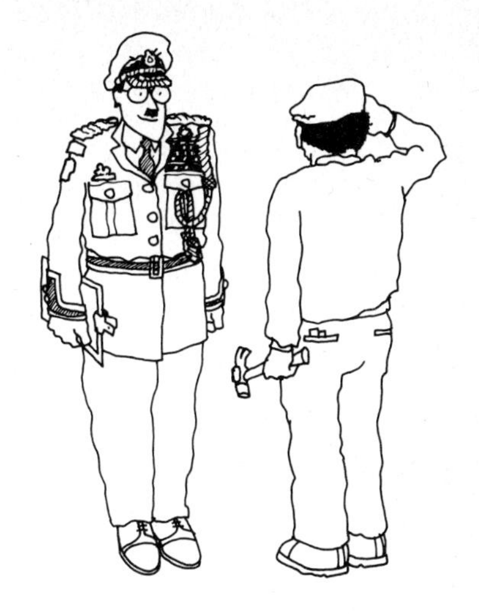

1 General

1.1 General notes

Location: Throughout building and site.

1 Check Dimensions: The contractor is to check all dimensions on site, and advise the architect of any discrepancies, particularly before ordering standard doors, windows, etc.

2 Refer to Part II: Throughout the Part III, Schedule of Works, the contractor must pay particular attention to the specifications of standards of materials and workmanship in the Part II. The headings to each subclause, e.g. 'Brickwork' refer to the relevant sections of Part II.

2 This note is specific to the Authors' Part II. It can be altered if required.

3 Measured Allowances: In this specification allowances for plaster repairs, pointing and floor repairs are provisional and to be measured with the architect before covering up. In addition wherever the term 'Allow' is used, the item of work described is to be remeasured before covering up.

4 Order of Work: The work included in the section of Part III, Schedule of Works, entitled 'Investigations' *must* be carried out at the beginning of the contract. In addition paper must be stripped from walls and ceilings and plaster specified to be hacked off must be removed at an early stage. Roof works must not be delayed. Any new work that might be damaged by water penetration because of out of sequence working will not be valued.

1.2 Schedule notes

Location: Throughout building.

1 Internal Door Schedule: For repairs to existing and supply and fixing of new internal doors, refer to the Internal Door Schedule. Price individually on the schedule, sum up and carry to Summary Sheet. Price the 'rates' column on the schedule.

2 External Door Schedule: For repairs to and supply of new windows and external doors, refer to the Window Schedule and External Door Schedule. Price individually on the schedules, sum up and carry to Summary Sheet. Price the 'rates' column on the schedules.

3 Window Schedule: For repairs to and supply of new windows and external doors, refer to the Window Schedule and External Door Schedule. Price individually on the schedules, sum up and carry to Summary Sheet. Price the 'rates' column on the schedules.

4 Ironmongery Schedule: For Ironmongery location, supply and fixing, or for fix only as required on the Schedule. Price only and carry to summary sheet.

5 Plaster Repair Schedule: For all plaster repairs (excludes plaster on *new* walls, partitions and ceilings) refer to Plaster Repair Schedule, price the schedule, sum up and carry to Summary Sheet. Price the 'rates' column on the schedule.

6 Internal Decoration Schedule: For internal decorations refer to Internal Decoration Schedule. Price the schedule and carry to Summary Sheet. Price the 'rates' column on the schedule.

7 External Decoration Schedule: For external decorations refer to External Decoration Schedule. Price the schedule and carry to Summary Sheet. Price the 'rates' column on the schedule.

8 Floor Finishes Schedule: For floor finishes refer to Floor Finishes Schedule. Price the schedule and carry to Summary Sheet. Price the 'rates' column on the schedule.

9 Kitchen Unit Schedule: For fix only of kitchen units refer to Kitchen Unit Schedule and P.C. Sum on Summary Sheet.

10 Sanitary Fitting Schedule: For fix only of sanitary fittings refer to Sanitary Fitting Schedule and P.C. Sum on Summary Sheet.

2 Protection/ Demolition/ Preliminary Investigations

2 Protection/ Demolition/ Preliminary Investigations

2.1 Protection/1

General

The aim of the demolition section is to obtain a 'cleaned out' structure, at which stage the provisional allowances for hidden work and any new problem not anticipated can be discussed with the engineer, QS, if any, and further instructions issued to the builder.

Make demolition descriptions accurate so that the extent is defined and measureable for the estimator. Take care not to include items of taking down specified in composite clauses within other parts of the Schedule of Works.

2.1.1 Protection

Location: ....

1 Item: cover the treads (and risers) of the staircase, with 4 mm hardboard pinned to existing. Protect balusters with 6 mm hardboard sheeting, wrap handrail in hessian strip.

1 Consider putting all items of general protection in the preliminaries. This clause should itemise particular features to be retained in situ and it may be necessary to specify how that is to be done. This is an example of one item. Further items can be either added to this clause or written in a new clause.

2.2 Demolition

2.2.1 Preliminary Stripping Out

1 Demolition: Instruct service authorities to disconnect gas and electricity supplies.

1 The provision of builder's supplies of gas, water and electricity can be dealt with in the contract preliminaries.

This clause may be deleted in many cases where the disconnection has already been made by the client or others.

2 Demolition: Strip out and remove all rubbish from the house, together with all redundant furnishings, fittings, finishes and services. Strip out with care to reduce making good to a minimum.

2 Exceptional quantities of rubbish should be mentioned specifically. For example if the house is full of furniture.

3 Demolition: Strip out and clear all rubbish from the site and the gardens.

3 Exceptional quantities of garden rubbish should be mentioned specifically.

2.2.2 General demolition

1 Demolition: Cut away, demolish and /or strip out all redundant structures, partitions and fixtures, including where shown on drawings: fireplaces, partitions, concrete floors, timber floors and associated skirting and joinery.

1 Care should be taken that particular demolitions where propping, etc. may be required, are only described in relation to the work required. These demolitions appear in later clauses.

The contractor is to inform the Local Authority Inspector before commencing any work which affects the structure of the building and make safe all temporary works.

2 Demolition: Strip out the following:
1.
2.
3.

2 There may be exceptional structures like brick garden sheds, garages or bomb shelters which should be specifically described as they may involve special plant and more labour than normal.

2.2.3 Setting aside (reuse on job)

1 Demolition: Take down, set aside and store for reuse on the job the following items:
1.
2.
3.

1 List here particular items. It may be necessary to say where they are to be stored. If they are too bulky or delicate provision should be made for removal from the building. Covering and protecting or special storage may be required in exceptional cases.

Ensure such items are clearly listed in Schedule of Conditions when the site is handed over.

2.2.4 Setting aside (collection by employer)

1 Demolition: Take down, set aside the following items for collection by employer and provide all labours for loading onto employer's vehicle:
1.
2.
3.

1 List here the items to be set aside and any further specific removal instructions.

2.3 Preliminary investigations

General

Only some of these items will be required—delete those not required. Particular items of inspection should be described in a separate clause. For example opening up brickwork to expose beams or concealed timber.

2.3.1 Structural inspection

1 Item: At the commencement of the job carry out the following works to permit Supervising Officer to inspect structure. Attend on Supervising Officer during inspection including provision of electric drill and 12 mm bit for testing timber lintels, etc.

2 Demolition: Lift one row of floorboards adjacent to all external walls on all suspended floors to permit examination of joist ends and wall plates. Similarly lift boards adjacent to all walls, including over sleeper walls, where suspended timber ground/basement floors are to be retained. Relay on completion.

3 Demolition: Cut away wall/ceiling plaster over upper floor of bay windows as directed to permit examination of bressummers.

3 This may not be necessary if the bay is to be re-roofed.

4 Demolition: Cut away wall plaster at ends of all window/external door lintels.

5 Excavation: Excavate .... no. trial holes size .... x .... adjacent to footings where directed by Supervising Officer.

5 Insert the number of trial holes and approximate width and depth required.

6 Item: Fix glass tell-tales across cracks in brickwork where directed by Supervising Officer. Tell-tales fixed with Epoxy Resin glue. Remove on completion. Immediately inform Supervising Officer of date tell-tales fixed.

6 Engineer's advice may have been taken on the structure where tell tales are needed; alter this clause to suit the type of observations required by the specialist.
Proprietory calibrated tell-tales are available if it is wished to monitor movement accurately.

2.4 Builder's work

2.4.1 Builder's work in connection with timber treatment

Location: ....

1 Carpentry: In addition to floorboards mentioned elsewhere, lift every 5th floorboard, lay to one side for spraying the floor cavity, re-lay in original position.

1 Allowance should be made under the floorboard section for replacement of a percentage of floorboards which will be damaged when removed for timber treatment.

2 Demolition: Expose all bonding timbers for application of timber treatment paste, if not directed to be exposed elsewhere in this specification.

2–5 Delete those clauses not required.

3 Plaster: Make good plaster in mix directed by Supervising Officer.
(To be measured in plaster repairs.)

4 Joinery: Remove box frames and set aside for inspection by Supervising Officer prior to timber treatment sub-contractor's arrival on site, if no other work specified in Window Schedule, replace box frames and make good finishes after timber has been treated. Window nos ....

4 Before using this clause decide which windows must be removed complete and be replaced and insert their numbers.

5 Joinery: Remove all shutter enclosures, timber panelling, and other timber trims forming cavities on to external walls; set aside for treatment by specialist subcontractor; replace and make good finishes.

3 Drainage

3 Drainage

General

Care must be taken not to undermine the existing structure with all new drainage work.

3.1 New underground drain runs

3.1.1 Test drains at commencement of construction

Location: ....

1 Drainage: Carry out approved tests on all retained existing drains at the commencement of construction and report results to Supervising Officer.

General

This clause should be used if the Supervising Officer has not tested the drains himself. Air pressure tests can 'blow out' old clay joints.

3.1.2 Locate existing drains

Location: ....

1 Excavation: Allow for the following excavation to locate exact run of existing drain runs: 2 no. holes 750 × 750 × 1000 mm deep.

1 This is necessary when the exact location of drain runs is unclear. It should be noted that the original drawings held by the Local Authority are not always accurate.

3.1.3 Line existing drain runs

Location: ....

1 Drainage: Provide the Provisional Sum of £ .... for lining of existing drain runs by specialist nominated subcontractor. Add for profit and attendance.

1 Insert a Provisional Sum based on quotations from a specialist subcontractor.

3.1.4 Grub up obsolete drainage

Location: ....

1 Excavation: Grub up obsolete drain runs and/or manholes as shown on drawings and backfill with compacted earth.

1 Some believe this to be an excessive specification and prefer merely to block off the ends of unused drain runs with concrete.

3.1.5 Lintels and reinforcement over drain runs

Location: ....

1 Brickwork: Insert 150 mm deep × (thickness of wall) PC RC lintels over all openings cut in walls for new drain runs.
Allow: .... no. lintels 600 mm long.

1 See Appendix for Standard lintels. Insert the number of lintels to be used.

2. Reinforcement: Reinforce all new concrete footings passing over new or existing drain runs with 3 no. 12 mm diameter high yield deformed bars. 50 mm polystyrene packing between top of drain run and concrete. Allow: .... no. sets of reinforcing bars 600 mm long.

2 This is for walls up to 450 mm thick. The materials and workmanship clauses should describe placement and cover of reinforcement. Insert the number of sets of reinforcement.

3.1.6 New underground drain runs (clay)

Location: ....

1 Excavation: Excavate drain trenches not exceeding 1000 mm deep and average 750 mm deep. Part return, fill and ram except where directed to backfill in concrete. Do not undermine adjacent foundations.

Option 1 Excavation: Excavate drain trenches not exceeding 2000 mm deep and average 1750 mm deep. Part return, fill and ram except where directed to backfill in concrete.

1 Select either 1 or option 1.
Remeasure actual excavation on completion.

2 Drainage: 100 mm diam. vitrified clay pipes as shown on drawings including all junctions and

2 Check with Local Authority about the acceptability of proprietory polystyrene joints.

bends. Bend(s) at base of soil stack(s) to be long radius bends.

3 Concrete: Bed and surround drains in 150 mm Mix A concrete.

4 Drainage: Agree falls with Supervising Officer and Local Authority before proceeding.

4 CP 304 requires vertical distance between lowest branch connection and inlet of drain to be at least 460 mm for a 2 storey house and 760 mm for taller dwellings. Falls for 100 mm pipe are normally 1:40. Always clearly note on drawings which drains are new and which are existing.

3.1.7 Concrete backfill to drain trenches

Location: ....

1 Concrete: Mix A where depth of excavation is equal to or greater than distance from edge of adjacent footings to nearest side of trench, or elsewhere where required by Supervising Officer or Local Authority.
Allow: .... cu. metres.

1 Avoid linking backfill concrete to new or existing. If they are linked the effective width of the footing will be increased which may bear on the drain.
Fill in number of cubic metres.

3.1.8 New underground drain runs (PVC)

General

Location: ....

Check with Local Authority for approval of PVC drains.

1 Excavation: Excavate drain trenches not exceeding 1500 mm deep and average 1000 mm deep. Part return in 300 mm layers compacted by hand.

1 Select either 1 or option 1. Remeasure upon completion if excavation varies by more than 300 mm.

Option 1 Excavation: Excavate drain trenches not exceeding 3000 mm deep and average 1750 mm deep. Part return in 300 mm layers, each compacted.

2 Drainage: 5–10 mm gravel bedding under drain. Width of trench × minimum 100 mm deep: following satisfactory drains test side fill, and backfill, full trench width × minimum 300 mm over drain.

2 Concrete backfilling is not used as it does not allow pipe to flex.
Depth of bedding will vary with soil condition.

3 Drainage: 110 mm UPVC pipes as shown, including all junctions and bends. Bends at base of soil stacks to be long radius.

3 For 2 storey house use 87½% long radius bend, 3 storey and over, use 2 no. 45° long radius bends.

4 Item: Agree falls with Supervising Officer and Local Authority before proceeding.

3.2 Gulleys/group connectors/reducing pieces

3.2.1 Gulley to new branch drain

Location: ....

1 Drainage: .... inlet gulley with CI grating connected to new branch.

1 Insert back or side.

3.2.2 Yard gulley to new branch

Location: ....

1 Drainage: Yard gulley with CI grating connected to new branch.

3.2.3 Group connector/reducing piece to new branch

Location: ....

1 Drainage: .... connected to new branch. See drawings for number of inlets.

1 Insert group connector or reducing piece.

3.2.4 Gulley to existing branch

Location: ....

1 Drainage: .... inlet gulley with CI grating connected to existing drain including breaking into existing drain and/or breaking out existing gulley.

1 Insert back or side.

3.2.5 Yard gulley to existing branch

Location: ....

1 Drainage: Yard gulley with CI grating connected to existing drain including breaking into existing drain and/or breaking out existing gulley.

3.2.6 Group connector/reducing piece to existing branch

Location: ....

1 Drainage: .... connected to existing branch including breaking into existing drain and/or breaking out existing gulley. See drawing for number of inlets.

1 Insert group connector or reducing piece.

3.2.7 Gulley kerb to existing gulley

Location: ....

1 Drainage: PC concrete gulley kerb bedded in position.

2 Drainage: CI grate with pre-formed opening to receive pipe connection.

General

This clause can be appropriate when retaining an existing gulley for a waste connection. Check that it is acceptable to the Local Authority.

3.2.8 Gulley to new branch (PVC)

Location: ....

1 Drainage: PVC trapped gulley and clip in grating, with rodding access and access point frame and cover, connected to new PVC branch.

2 Concrete: Backfill in Mix A to top of vertical socket, backfill and compact in 'as dug' material.

1 Gullies are available with built-in rodding access, a separate access may not then be required.

3.2.9 Vertical inlet gulley to new branch (PVC)

Location: ....

1 Drainage: PVC trapped gulley with vertical inlet hopper and clip in grating, cut to suit, with rodding access, access point frame and cover, connect to new PVC branch.

2 Concrete: Backfill in Mix A to top of vertical socket, remainder backfill and compact in 'as-dug' material.

3.2.10 Back inlet gulley to new branch (PVC)

Location: ....

1 Drainage: PVC trapped gulley with clip in grating, with horizontal inlet piece, and rodding access, access point frame and cover; connected to new PVC branch.

2 Concrete: Backfill in Mix A to top of vertical socket, remainder backfill and compact in 'as dug' material.

3.2.11 Rodding eye to new PVC drains

Location: ....

1 Drainage: PVC 'seated' drain rodding access, up to maximum invert level of 815 mm, medium radius bend, extension piece, screwed access cover, rodding access point frame and cover.

3.3 Manholes

General

Check whether sulphate resisting cement is required—if ground is normally wet then it may be required—check with the Local Authority. Take care that adjacent foundations are not disturbed.

3.3.1 New manhole (up to 1000 mm deep)

Location: ....

1 Excavation: To required extent to give invert level approx .... mm. Part return, fill and ram except where directed to backfill in concrete.

1 Insert invert level. Remeasure depth on completion.

2 Concrete: 150 mm thick base slab Mix B finished level with external face of brickwork, Size .... × ....

2 Insert dimension. Length of manhole as follows: 1030 mm allows 2 to 3 no. branches each side. Additional branches require 250 mm more each ideally (but this can be compressed).

3 Brickwork: Semi-engineering bricks one brick thick in 1:3 mortar. English bond flush joint. Corbel sides in 58 mm steps to required size for cover frame. Internal dimensions 658 mm × .... mm.

3 Insert dimension.

4 Drainage: Half-round main channel connected to main drain run with drain chutes (or intercepting trap if required by Local Authority). Branch and branch channels as shown on drawings to pick up branch drains. Mix B concrete benching.

4 Select either 4 or option 4.

Option 4 Drainage: Proprietary inspection chamber installed to manufacturer's instructions.
Manufacturer ....
Cat/Type no. ....

Option 4 Insert manufacturer and Type number.

5 Drainage: 610 × 457 mm CI single seal cover and frame.

Option 5 Drainage: 610 × 457 mm CI double seal double cover and frame (locking 4 no. screws).

5 Select either 5 or option 5. Option gives internal manhole. If internal manhole is set below timber floor a double seal double cover is not required by all Local Authorities but is advisable. Allow elsewhere in specification for hatch in timber floor.

6 Item: The contractor is to insert his rate, plus or minus, for each 300 mm of depth.
300 mm @ £ ....

3.3.2 New manhole (over 1000 mm deep)

General

A specially written clause will be required for exceptionally deep manholes.

Location: ....

1 Excavation: To required extent to give invert level approx .... mm. Part return, fill and ram except where directed to backfill in concrete. Prop and shore as required.

1 Insert invert level. Remeasure depth upon completion.

2 Concrete: 230 mm thick base slab Mix B finished level with external face of brickwork.

3 Brickwork: Engineering brick walls one brick thick in 1:3 mortar. English bond flush joint. Internal dimensions 685 × .... mm. Step irons built in every fourth course set staggered in two vertical runs 230 mm apart.

3 Insert dimensions. Length of manhole as follows: 1030 mm allows 2 to 3 no. branches each side. Additional branches require 250 mm more each ideally (but this can be compressed).

4 Reinforcement: Square mesh reinforcement to BS 4483 Code A142, 200 × 200 mm, 2.2 kg/sq.m in cover slab.
Minimum cover to reinforcement 40 mm.

4 and 5 Suitable for deep manholes. Can corbel as **3.3.**1.

5 Concrete: Mix B reinforced slab 100 mm thick with opening for manhole cover.

6 Drainage: Half-round main channel connected to main drain run with drain chutes (or intercepting trap if required by Local Authority). Benching and branch channels as shown on drawings to pick up branch drains. Mix B concrete benching.

7 Drainage: 610 × 457 mm CI single seal cover and frame.

7. Select either 7 or option 7.

Option 7 Drainage: 610 × 457 mm CI double seal double cover and frame (locking 4 no. screws).

Option 7 gives internal manhole. If internal manhole is set below timber floor a double seal double cover is not required by all Local Authorities but is advisable. Allow elsewhere in specification for hatch in timber floor.

8 Item: The contractor is to insert his rate, plus or minus, for each 300 mm of depth.
300 mm @ £ ....

3.3.3 Tumbling bay

Location: ....

1 Drainage: Extra over Clause .... for 100 mm diameter cast iron tumbling bay with inspection eye connected to clay drain at high level and discharging over manhole benching.

1 Insert Clause number.

3.3.4 Interceptor pull

Location: ....

1 Item: Provide length of 6 mm polypropylene rope knotted to grease interceptor plug in manhole and looped over galvanised staple fixed to brickwork at top of manhole.

1 May be required in existing manholes leading to the sewer, as well as new manholes.

3.4 Soil and waste stacks

3.4.1 New SVP

Location: ....

General

The following clauses assume a single stack system.

1 Drainage: 100 mm .... PVC soil stack to accept branches shown. Connect to underground drain at base and terminate 900 mm above head of highest window with PVC terminal. Plastic plug and screw to brickwork using sheradised screws. Ensure that allowance is made for movement joints.

1 Insert either grey, black or other colour that is available.

2 Drainage: Extra over above for bottom 1800 mm of stack in cast iron.

2 Include this clause if required by Local Authority or if in a position where it could be damaged.

3 Roofing: Proprietary neoprene sleeved weathered slate where stack passes through roof finish.

3 Include this clause only if the SVP penetrates the roof.

4 Drainage: Access door at base of stack and at each WC connection, in agreed positions.

3.4.2 New waste stack

Location: . . . .

1 Drainage: 50 mm . . . . PVC waste stack to accept branches shown. Connect to underground drain /gulley at base and terminate 900 mm above head of highest window with PVC terminal. Plastic plug and screw to brickwork using sheradised screws.

1 Insert either grey, black or other colour that is available.

2 Drainage: Extra over above for bottom 1800 mm of stack in cast iron.

2 Include this clause if required by Local Authority or if in a position where it could be damaged.

3 Roofing: Provide proprietary neoprene sleeved weathered slate where stack passes through roof finish.

3 Include this clause only if the SVP penetrates the roof.

3.4.3 Stub soil stack

Location: . . . .

1 Drainage: 100 mm . . . . PVC stub soil stack with screwed capped end and to accept branches shown. Connect to underground drain at base. Overall height of stack approximately 900 mm.

1 Insert either grey, black or other colour that is available.
This arrangement permits ground floor bath, basin and W.C. etc. to be connected together, thus possibly saving several underground drain runs.
If stub stack is in a position where it may be damaged it should be in cast iron.

3.5 Waste and soil branches/traps

3.5.1 Branch/wastes/traps

Location: . . . .

1 Drainage: Polypropylene push fit branch wastes as shown connected separately to stack. Do not connect wastes less than 200 mm below W.C. connections. Provide access caps at every change of direction in waste branches except shallow sweep bends. Sizes of wastes as follows:

WCs	110 mm
Baths	40 mm
Sinks	40 mm
Washing machines	40 mm
Basins	32 mm

Option 1 Drainage: High temperature PVC solvent weld branch wastes as shown connected separately to stack. Do not connect wastes less than 200 mm below W.C. connections. Provide access caps at every change of direction in waste branches except shallow sweep bends. Sizes of waste as follows:

WCs	110 mm
Baths	40 mm
Sinks	40 mm
Washing machines	40 mm
Basins	32 mm

1 Select either 1 or option 1.
BS 5572: 1978 gives following maximum lengths for waste connections to single stack system for waste sizes as above:

Sink	*2.3 m*	*3.0*
Basin	*1.7 m*	*3.0 (with 40 mm discharge)*
Bath	*2.3 m*	*3.0*
WC	*1.5 m*	*6.0 (for straight run)*

Push fit waste systems have the advantage that you can dismantle them if there is a blockage, but if the run is exposed they could be damaged by accident, it is important with solvent weld systems that access is allowed at any significant change of direction and at each trap, and to allow for movement.
Check that the sizes of pipes given in this clause are sufficient for the distance between the fitting and the stack and amend as necessary.
BS 5572 allows use of anti-syphon devices.

2 Drainage: 75 mm deep seal 'P' or 'S' traps to baths and 75 mm deep seal bottle traps to basins and sinks.

3 Drainage: Washing machine standpipe with screw capped end clipped to wall at height of 800 mm above FFL 75 mm deep seal 'P' or 'S' trap.

3 Include only if required.
Waste branch pipes are available in black. Specify change of colour on face of building if required.

3.6 Rainwater pipes and gutters

3.6.1 Rainwater pipes and gutters

General

Houses with very large roofs may require larger capacity gutters and downpipes.

Location: ....

1 Drainage: 100 mm .... half round PVC gutters and brackets fixed with sheradised screws to fascias at all eaves complete with all outlets, stop ends and fittings, including stop end to neighbour's gutter(s) where appropriate.

1 Insert black, grey or other available colour.

2 Drainage: 62 mm .... PVC down pipe(s) as shown, complete with branch(es) and shoe(s). Plastic plug and screw to brickwork using sheradised screws.

2 Insert black, grey or other available colour. Insert square or round.

3 Drainage: Extra over above for bottom 1800 mm of stack in cast iron.

3 Include this clause if required by Local Authority or if in a position where it could be damaged.

3.6.2 Rainwater pipe and hopper

General

Houses with very large roofs may require larger capacity gutters and downpipes.

Location: ....

1 Drainage: 75 mm .... PVC down pipe(s) as shown complete with hopper head(s), branch(es) and shoe(s). Plastic plug and screw to brickwork using sheradised screws.

1 Insert black, grey or other available colour.

2 Drainage: Extra over above for bottom 1800 mm of stack in cast iron.

2 Include this clause if required by Local Authority or if in a position where it could be damaged.

3.7 Testing

3.7.1 Completion

1 Drainage: Rod through. Leave all drains, gulleys and manholes clean and clear on completion with manhole covers bedded in grease.

4 External works

4.1 Concrete Work
4.1.1 Concrete garden steps

4.2 Brickwork
4.2.1 Brick retaining wall (maximum height 900 mm)
4.2.2 Brick retaining wall (maximum height retained 1350 mm)
4.2.3 One brick thick garden wall
4.2.4 Alter existing brick garden wall

4.3 Timber fencing
4.3.1 Paling fence (for painting)
4.3.2 Paling gate (for painting)
4.3.3 Close boarded fence
4.3.4 Close boarded gate
4.3.5 Paling fence (sawn SW for creosote)
4.3.6 Repairs to existing fence
4.3.7 Chain link fence

4.4 Metal railings
4.4.1 Repairs to metal railings (cast iron)
4.4.2 New metal railing
4.4.3 New metal handrail to steps
4.4.4 New metal gate

4.5 Pavings/footings for metal or timber structures
4.5.1 Paving concrete
4.5.2 Paving slabs
4.5.3 Footing for straight stair
4.5.4 Base for spiral stair

4.6 Timber external stair
4.6.1 Timber stair

4.7 Metal external stairs
4.7.1 Metal spiral stair
4.7.2 Metal stair

4.8 Clearing gardens/digging over/planting
4.8.1 Prepare garden
4.8.2 Seed to grass
4.8.3 Plant tree(s)

4 External works

4.1 Concrete work

4.1.1 Concrete garden steps

Location: ....

1 Excavation: To suit rake of steps as shown with 450 mm × 450 mm trench at base.

1 This clause assumes that excavation reveals solid ground. Depending on subsoil, hardcore may be required.

2 Formwork: To suit.

3 Concrete: Mix B steps with minimum waist of 150 mm .... mm treads × .... mm risers as shown. Exposed surface to be fairfaced.

3 Insert dimensions of treads and risers.

4 Cross refer: Refer to drawing no ....

4 Insert drawing number. A drawing should be done unless there are few steps.

4.2 Brickwork

4.2.1 Brick retaining wall
(Maximum height 900 mm)

Location: ....

1 Excavation: To extent shown on drawings and for foundation trench 1000 mm deep and (0.67 × height of retained earth) wide.

2 Concrete: Mix A trenchfill footing 800 mm × 600 mm wide.

3 Brickwork: One brick wall. Well burnt second hand facing bricks. Flettons in backings. English bond. Brick on edge coping with galvanised ms cramps at exposed ends. Build in 38 mm diameter PVC weepholes at 900 mm centres, 150 mm above lower ground level. Overall height .... mm above higher ground level. Flush joints where exposed.

3 Insert height. Other facing bricks and joints can be used. Consider water drainage from the higher level.

4 Hardfill: Backfill on completion with hardcore. Minimum 250 mm thick behind retaining wall.

4 Hardcore: Backfill reduces lateral pressure and damage from clay swell.

5 Brickwork: Insert vertical leadcore DPC to all positions where retaining wall abuts external walls of house from below main wall dpc to top of retaining wall.

4.2.2 Brick retaining wall
(Maximum height retained 1350 mm)

Higher retaining walls should be designed for the specific situation.

Location: ....

1 Excavation: To extent shown on drawings and for foundation trench 1000 mm deep and (0.67 × height of retained earth) wide.

2 Concrete: Mix A trenchfill footing 800 mm deep × 600 mm wide.

3 Brickwork: One and a half brick wall. Well burnt second hand facing bricks. Flettons in backings. English bond. Brick on edge coping with galvanised ms cramps at exposed ends. Build in 38 mm diameter PVC weepholes at 900 mm centres 150

3 Insert height. Other facing bricks and joints can be used.
Waterproof render to back face and land drainage may be required to avoid salts and water penetration.

mm above lower ground level. Overall Height .... mm above higher ground level. Flush joints where exposed.

4 Hardfill: Backfill on completion with hardcore. Minimum 250 mm thick behind retaining wall.

4 Select either 4 or option 4. See note previous clause.

Option 4 Hardfill: Backfill on completion in Mix A concrete.

Option 4 Concrete backfill should be considered if the retained material requires consolidation.

5 Brickwork: Insert vertical leadcored dpc at all positions where retaining wall abuts external walls of house, from below main wall dpc to top of wall.

6 Cross Refer: Refer to drawing no ....

6 Insert drawing no. A drawing may be necessary to show the relation of the new and existing footings. A movement joint may be required.

7 Return 300 mm top soil to levels required.

4.2.3 One brick thick garden wall

Location: ....

1 Excavation: For foundation trench, 600 mm deep. Backfill topsoil on completion.

2 Concrete: Mix A footing in trench 300 mm deep × 450 mm wide.

2 For lengths greater than 5000 mm strip footings should be broken at expansion joints with 12 mm gap.

3 Brickwork: One brick thick wall. Second hand stock facing bricks. English garden wall bond with flush joints all around. Brick on edge coping with galvanised ms cramps at exposed ends. Overall height generally .... mm and as shown. Length .... mm 12 mm open expansion joints with vertical leadcored dpc at abutment with wall of house.

3 Insert height and length. Other facing bricks and joints can be used. Maximum height for this wall is 1800 mm. English garden wall bond is one complete course of headers to 3 to 5 courses of stretchers.

4 Brickwork: One and a half brick square piers at same centres as existing (but not greater than 3000 mm centres) and on each side of existing expansion joint. 12 mm open expansion joints every 6000 mm.

4 This clause should be included for longer walls where piers and expansion joints are required.

5 Brickwork: Brick bond into adjacent existing brick garden wall or walls.

5 Select this clause if appropriate. Movement joint may be required.

6 Brickwork: 2 courses of tile creasings under brick on edge coping.

6 Select either 6 or option 6 or write a new clause which allows for another alternative such as a pre-cast concrete or stone coping.

Option 6 Brickwork: Brick sailing course under brick on edge coping.

4.2.4 Alter existing brick garden wall

Location: ....

1 Demolition: Take down .... courses of brickwork and/or copings .... long and set aside for reuse.

1 Insert number of courses and length.

2 Brickwork: In position shown on drawings build up on existing brick base in bricks to match existing in English garden wall bond with flush joints all around brick on edge coping with galvanised ms cramps at all exposed ends. Overall height generally .... mm and as shown. Length .... mm and as shown.

2 Insert height and length.

3 Brickwork: Brick bond into adjacent brick garden wall or walls.

4 Brickwork: 2 courses of tile creasings under brick on edge coping.

Option 4 Brickwork: Brick sailing course under brick on edge coping.

5 Pointing: Rake out existing brickwork to a depth of 20 mm and point with a flush joint. Allow .... sq m.

6 Cross Refer: Refer to drawing no ....

4 Select either 4 or option 4 or write a new clause 4, which allows for another alternative such as precast concrete or stone coping.

5 Include if required and insert the area.

6 A detail drawing may be required.

4.3 Timber fencing

General

Complete run of fencing must be indicated on the site plan.

4.3.1 Paling fence (for painting)

Location: ....

1 Excavation: Holes, 600 mm deep for post surrounds. Backfill on completion.

2 Concrete: Mix A post surrounds 300 × 300 × 400 mm deep.

3 Carpentry: Bed 100 × 100 mm wrot tanalised sw posts with twice weathered tops at 2700 mm centres. Ex 25 × 100 mm wrot top and bottom rails housed to posts. Ex 75 × 19 mm wrot palings with twice weathered tops fixed to rails at 150 mm centres with galvanised nails. Overall height .... mm. Allow .... linear metres.

3 Insert height.
Insert length.

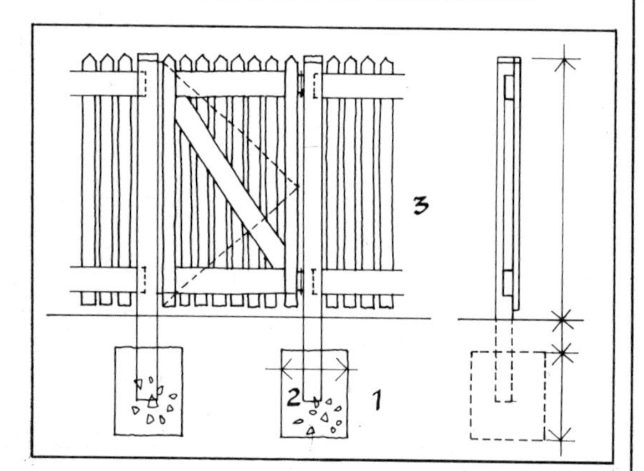

4.3.2 Paling gate (for painting)

Location: ....

1 Excavation: Holes 600 mm deep for post surrounds.

2 Concrete: Mix A post surrounds 300 × 300 × 400 mm deep.

3 Carpentry: Bed ex 100 × 100 mm sawn tanalised posts with twice weathered tops to receive gate hinges and latch. Post ends dipped in liquid bitumen before fixing. Posts to same height as fence.

4 Carpentry: Ex 25 × 100 mm wrot sw rails and diagonal brace and ex 75 × 19 mm wrot sw palings with twice weathered tops *screwed with sheradised screws.*

5 Carpentry: Hang gate in 2 no. galvanised iron 'T' hinges, size .... long, with one no. spring latch.

5 Insert hinge length.

4.3.3 Close boarded fence	General
Location:	*Complete run of fencing must be indicated on the site plan.*
1 Excavation: Holes 600 mm deep for post surrounds.	
2 Concrete: Mix A post surrounds 300 × 300 × 400 mm deep.	
3 Concrete: Bed precast morticed concrete posts to BS 1722 Part 5, at 2700 mm centres to give overall fence height of .... mm.	*3 Insert height—standard fence boarding heights are 1500 mm and 1750 mm.*
4 Carpentry: House 2 no. ex 75 × 75 mm triangular aris rails to posts. Fix 100 mm feather edged boards with galvanised nails. Fix 150 × 25 mm sw gravel board. All timber pressure creosoted to BS 913. Steep all cut ends in creosote for 5 minutes.	
5 Allow: .... linear metres.	*5 Insert length.*
4.3.4 Close boarded gate	
Location:	
1 Excavation: Holes 600 mm deep for post surrounds.	
2 Concrete: Mix A post surrounds to 300 × 300 × 400 mm deep.	
3 Carpentry: Bed in concrete 100 × 100 mm tanalised sw posts with twice weathered tops. Dip ends in liquid bitumen before fixing, to same height as fencing.	
4 Carpentry: 100 × 25 mm rails and diagonals. Feather edge boards nailed with galvanised nails.	
5 Carpentry: Hang gate on 2 no. galvanised iron 'T' hinges, size .... mm long, with one no. spring catch.	*5 Insert hinge length.*
4.3.5 Paling fence (sawn sw for creosote)	General
Location:	*Complete run of fencing must be indicated on the site plan.*
1 Excavation: Holes 600 mm deep for post surrounds. Backfill on completion.	
2 Concrete: Mix A post surrounds 300 × 300 × 400 mm deep.	
3 Carpentry: Bed 100 × 100 mm sawn sw posts, with twice weathered tops at 2700 mm centres. House 25 × 100 mm sw top and bottom rails to posts. Fix 75 × 19 mm sawn sw palings with twice weathered tops to rails at 150 mm centres with galvanised nails. Overall fence height .... mm. All timber pressure creosoted to BS 913. Steep all cut ends in creosote for five minutes.	*3 Insert height.*
4 Allow: .... linear metres.	*4 Insert length.*
4.3.6 Repairs to existing fence	
Location:	
1 Demolition: Take down existing fence, setting aside rails, posts and boarding.	*1 This is to allow for the reuse of sound timber.*

2 Carpentry: Cut off all rotten timber from posts, and agree work to remaining timbers with Supervising Officer. Steep all cut ends in creosote for 5 minutes.

3 Carpentry: Rebuild existing fence, sitting posts in proprietary metal post sockets. Fix new gravel board 150 × 25 mm sw. All new timbers to be finished in 2 coats of creosote. Sockets to be ....

3 Insert manufacturer and type.

4 Allow: .... linear metres of repairs.

4 Insert length.

4.3.7 Chain link fence

Location: ....

1 Excavation: Holes 600 mm deep for post surrounds. Backfill with compacted topsoil over surrounds.

2 Concrete: Mix A post surrounds 300 × 300 × 400 mm deep for intermediate posts, 450 × 450 × 400 mm for straining posts.

3 Item: PC reinforced concrete fencing posts at equal centres, maximum 3 m for intermediate posts, straining posts at all ends, corners, changes of direction, all to BS 1722 Part 1.

4 Item: .... mm high galvanised chain link fencing with galvanised straining wires, fittings and tie wires. Bottom of fence to be kept clear of earth. All to BS 1722 Part 1.

4 Insert height. The standard heights that are available are 900 mm, 1200 mm and 1400 mm. Plastic coated fence and wires are also available to the same BS if desired.

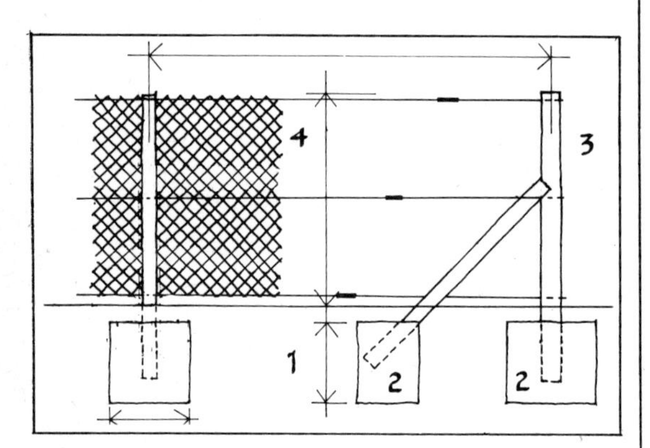

4.4 Metal railings

4.4.1 Repairs to metal railings (cast iron)

Location: ....

1 Metalwork: .... no. new or second hand stanchions to match existing, including heads.

1 Insert number of stanchions.

2 Metalwork: Clear out .... no. pockets to copings and bed stanchions in molten lead.

2 Insert number of pockets.

3 Concrete: Cut away existing coping and cast new concrete coping Mix A around railings to match existing profile. Top surface trowelled to fall 12 mm to edges.

3 This clause applies if the repaired coping is to be cast around the existing railing bases. If cost allows, polystyrene filled pockets should be formed first and the old railings replaced as 2 above.

4 Metalwork: Refix mild steel/cast iron rails to walls/stanchions.

4 This clause requires further detail to describe particular fixing repairs.

5 Allow: .... no. welds.
Allow: .... no. drilled and bolted connections with 150 × 38 × 9 mm ms flanges.

5 Insert number of welds and connections.

4.4.2 New metal railing

Location: ....

1 Brickwork: Drill pockets 250 mm deep into brickwork/stonework to receive stanchions. Form pockets for railing ends.

2 Metalwork: Metal railing finished height .... mm. 38 × 12.5 mm convex section ms handrail. 38 × 38 × 10 gauge square section tubular stanchions with sealed ends, at equal centres (maximum 1000 mm). 38 × 6 mm ms flat bottom rail positioned 100 mm above brickwork. 8 mm square ms balusters at equal centres (maximum 108 mm). Curl exposed ends of handrail and fishtail ends for building in.

2 Insert height.

3 Metalwork: Grout into pockets. Top 50 mm of pockets filled with molten lead.

4 Painting: 2 coats red oxide primer to railing before fixing.

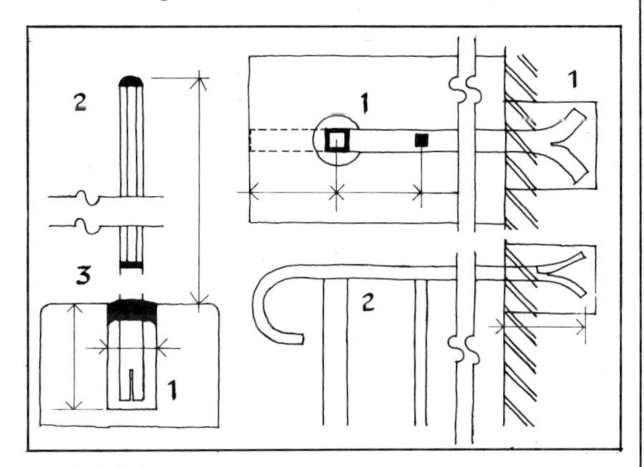

4.4.3 New metal handrail to steps

Location: ....

1 Brickwork: Drill pockets 250 mm deep into brickwork/stonework to receive railing.

2 Metalwork: Metal handrail, finished height .... mm. 25 mm internal diameter ms barrel formed to suit line of steps and bent to fit pockets at each end. Grout into pockets. Top 50 mm of pockets filled with molten lead.

2 Select either 2 or option 2. Insert height.

Option 2 Metalwork: Metal handrail, finished height .... mm. 25 mm internal diameter ms barrel formed to suit line of steps and bent to fit pockets each end with .... no. similar welded intermediate supports.

Option 2 Insert height and number of intermediate supports. Use this option if intermediate supports are required.

3 Metalwork: Grout into pockets. Top 50 mm of pockets filled with molten lead.

4 Painting: 2 coats red oxide primer to railing before fixing.

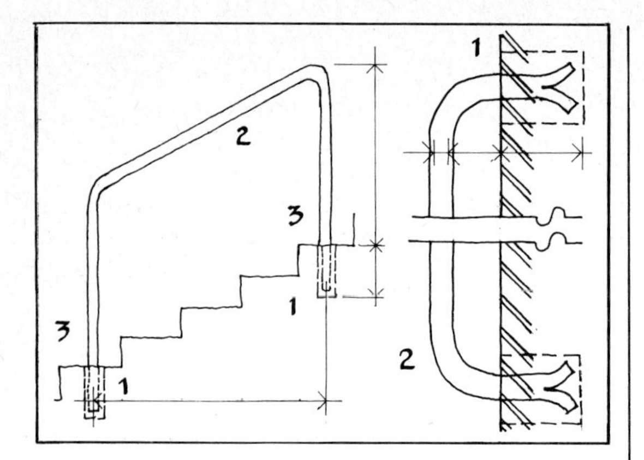

4.4.4 New metal gate

Location: ....

1 Excavation: To receive 300 × 300 × 400 mm holes for each post.

2 Metalwork: MS gate with 2 no. ms posts bedded in concrete.

2 Select either 2 or option 2.

Option 2 Metalwork: MS gate with 2 no. ms posts bedded in concrete type .... by

....

Option 2 Insert type and supplier.

3 Cross refer to drawing no. ....

3 Insert drawing number if appropriate.

4.5 Pavings/footings for metal or timber structures

General

Only a limited number of finishes have been included. There are many other options that could be used.

4.5.1 Paving concrete

General

Location: ....

The site plan should indicate the full extent of the paving.

1 Excavation: Excavate level base .... approx below existing ground level. Strip and set aside top soil for areas over 10 sq. metres.

1 Insert depth of excavation.

2 Hardfill: 100 mm hardcore bed and sand-blinding laid to falls away from house.

3 Concrete: Mix A slab 75 mm thick, laid to falls away from house. Finish to a level surface with a wood float. Score into 600 mm squares while still green. Finished paving level should be 150 mm below FFL adjacent, but do not expose footings to achieve this. Instead consult Supervising Officer before proceeding.

4.5.2 Paving slabs

General

Location: ....

The site plan should indicate the full extent of the paving.

1 Excavation: Excavate level base .... approx below existing ground level. Strip and set aside topsoil for areas over 10 sq. metres.

1 Insert depth of excavation.

2 Item: Minimum 40 mm 1:3 lime/sand bed laid to falls away from house. 600 × 600 × 50 mm PC concrete paving slabs to BS 368 well pressed into

bed. Joints filled with 1:3 lime/sand brushed in dry. All cutting by abrasive saw. Finished paving level should be 150 mm below FFL adjacent but *do not* expose footings to achieve this. Instead consult Supervising Officer before proceeding.

3 Item: Infill around paving slabs with washed gravel, to pass 12 mm sieve, to minimum depth 50 mm finished level with paving.

3 Include if appropriate. Cutting slabs is expensive and not usually done well. This can sometimes be avoided by using this clause or leaving a gap for a planting bed.

4.5.3 Footing for straight stair

Location: . . . .

General

This clause assumes a conventional straight single flight metal or timber stair with metal shoes to be rawlbolted to the concrete. The dimensions relate to 600 mm square paving slabs.

1 Excavation: For footing 450 mm deep. Backfill on completion.

2 Concrete: Mix B 1200 × 600 × 450 mm deep. Top level with adjacent paving. Finish to a level surface with a wood float.

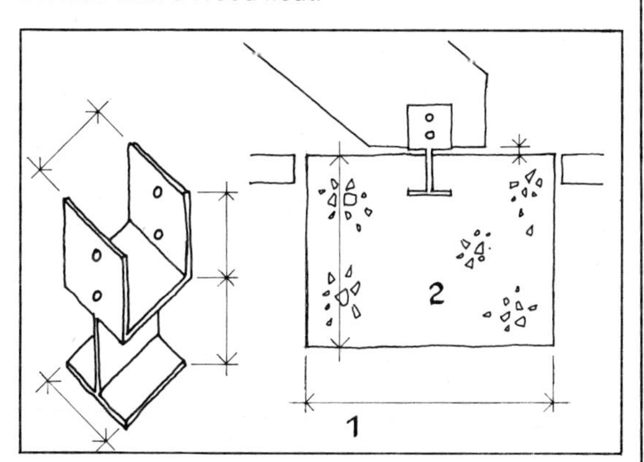

4.5.4 Base for spiral stair

Location: . . . .

1 Excavation: Excavate level base.

2 Hardfill: 100 mm hardcore bed and sand-blinding.

3 Concrete: Mix B slab . . . . mm thick . . . . mm × . . . . mm laid to fall away from house. Top to be level with adjacent paving, with a wood float finish. Score into 600 mm squares while still green to line with joints of any adjacent paving slabs.

3 Insert dimensions: 150 mm thick and 1200 mm × 1200 mm is adequate for most domestic spiral stairs. Check soil conditions to see if 150 mm is adequate in the particular circumstances.

4 Cross Refer: Refer to Drawing no. . . . . for stair fixings.

4 Insert drawing number if appropriate.

4.6 Timber external stair

4.6.1 Timber stair

Location: . . . .

1 Carpentry: Construct new timber stair as drawing no. . . . .

1 A detailed drawing is required to fully describe this stair.

4.7 Metal external stairs

4.7.1 Metal spiral stair

Location: ....

1 Metalwork: Galvanised pressed metal spiral stair and landing fixed in position as shown on drawings, balusters at maximum 100 mm clear cc's and bar across riser space. As shown on drawing no. .... and supplied by

....

1 Insert the drawing number and name of manufacturer.
A detailed drawing is necessary to describe fully the number of risers and landing condition, etc.

An alternative to this clause is to make the stair a nominated subcontract item as many manufacturers offer an erection service.

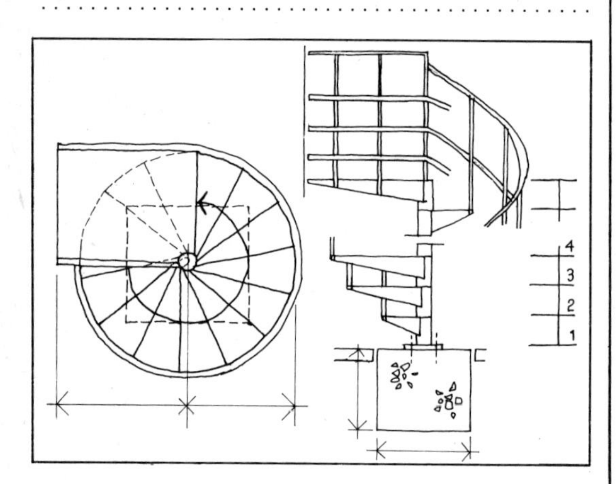

4.7.2 Metal stair

Location: ....

1 Metalwork: Metal staircase as drawing no. .... Fixed in position as shown on drawings. Provide Supervising Officer with shop drawings before fabrication commences.

1 Insert drawing number.
A detailed drawing is necessary to describe the stair fully. An alternative to this clause is to make the stair a nominated subcontract item.

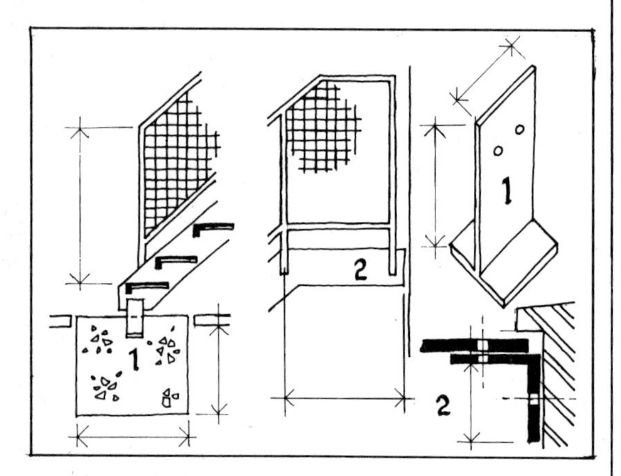

4.8 Clearing gardens/digging over/planting

4.8.1 Prepare garden

Location: ....

1 Item: Clear garden of all builder's rubbish, dig over, rake and consolidate to provide a level tilth suitable for seeding.

4.8.2 Seed to grass

Location: ....

1 Item: Sow with a general purpose lawn mixture at a rate of 36 grams/square metre ($1\frac{1}{2}$ oz/square yard). Lightly rake on completion.

General

Only use this clause if seed can be sown between April and September inclusive.

4.8.3 Plant tree(s)

Location: ....

1 Item: Plant .... no. tree(s) Species .... with a minimum clear height of 2000 mm with .... no. stakes and 2 no. ties per tree.

1 Insert number of trees and species and number of stakes and ties.

5 External and party walls

5.1 Chimney breast removal/alterations
5.1.1 Remove chimney breast

5.2 New external wall
5.2.1 New external solid wall
5.2.2 New cavity wall

5.3 Repair existing brickwork
5.3.1 Rebuild defective brickwork (facing bricks)
5.3.2 Rebuild defective brickwork (rendered work)
5.3.3 Cut out timber wall plates
5.3.4 Piece in facing bricks
5.3.5 Reinforced concrete corner ties
5.3.6 CP III lateral restraint straps (direction of joist span)
5.3.7 CP III lateral restraint straps (right angles to span)
5.3.8 Steel ties
5.3.9 Expanded metal mesh tie
5.3.10 Stitch cracks

5.4 Blocking openings
5.4.1 Block miscellaneous small openings
5.4.2 Block fireplace
5.4.3 Block fireplace (for gas fire)
5.4.4 Block fireplace (for back boiler)
5.4.5 Block flues (fireplace as recess)
5.4.6 Block external openings (with new footings)
5.4.7 Block external opening (on existing footings, or at high level)

5.5 Forming/altering openings
5.5.1 Form new opening/adjust existing opening
5.5.2 Raise cill height (with new footings)
5.5.3 Raise cill height (on existing footings or at high level)
5.5.4 Lower cill height
5.5.5 Replace existing lintels/brick arches
5.5.6 Replace concrete cill
5.5.7 Replace concrete cill (reinforced)
5.5.8 Brick up recessed reveals
5.5.9 Airbrick/airvent
5.5.10 Airbrick/airvent
5.5.11 Airbrick/airbrick
5.5.12 Airvent in chimney breast
5.5.13 Balanced flue boiler outlet

5.6 Parapets and stacks
5.6.1 Repair parapets
5.6.2 Copings (brick on creasing tile)
5.6.3 Copings (brick on sailing course)
5.6.4 Copings (concrete)
5.6.5 Copings (tile)
5.6.6 Rebuild stack (retaining pots)
5.6.7 Rebuild stack (removing pots)
5.6.8 Take down stack and cap off
5.6.9 Cap off existing pots
5.6.10 Remove existing pots

5.7 External wall finishes
5.7.1 Pointing
5.7.2 Rendering (replacing existing render)
5.7.3 Rendering (to existing facing brick)

5 External and party walls

5.1 Chimney breast removal/alterations

5.1.1 Remove chimney breast

Location: ....

General

1 This clause assumes the presence of a new concrete ground floor slab and total demolition of stack from roof to ground level. Together with stripping all roof coverings.

2 If stack is not shared then (generally) take down stack completely.

3 Remember that some breasts slim down on the first floor, and that flues of both houses are within the thickness of the party wall. Such breasts cannot be demolished, and must therefore be supported within the floor zone.

4 In London Building Act Amendment 1939 areas, apply for a Chimney Breast Certificate.

5 If stack must be kept for appearance sake then the method of support of the stub will require a detail.

1 Demolition: Work from top of stack down to ground level. Carefully demolish stack and breast flush with face of party wall leaving as many sound headers as possible for bonding with new work. Clear out all soot, pargetting and old mortar from exposed flues. Break out hearths complete.

2 Carpentry: Take up floorboards as necessary and set aside for reuse.

3 Brickwork: Build up exposed flues in flettons, bonding in with new or existing headers as frequently as possible.

4 Carpentry: Replace .... no. roof rafters to same dimensions as existing. Refix new joists with ....

4 Insert number of roof rafters and describe fixing method.

5 Carpentry: Replace .... no. ceiling rafters to same dimensions as existing. Refix new joists on ....

5 Insert number of ceiling joists and describe fixing method.

6 Carpentry: Remove trimmers to hearth and unsupported joists. Replace .... no. floor joists to same dimensions as existing. Refix new joist on ....

Select either 6 or option 6.
6 Assumes joists are at right angles to wall of chimney breast. Insert number of joists and describe fixing method.

Option 6 Carpentry: Tanalised sw trimmers 50 mm × (same depth as existing) to pick up unsupported intermediate floor joists. Fix to hearth trimmer, unsupported joists and party wall with joist hangers. 2 no. 50 mm × (same as depth of existing) tanalised stub joists spanning between trimmer, fixed to trimmers with joist hangers.

Option 6 Assumes joists are parallel to chimney breast wall.

7 Carpentry: New sw plain edge floor boards over openings in floors. Width, thickness and direction equal to existing adjacent, ends staggered. Re-lay boards set aside.

8 Render: Mix 1 to demolished face of stack. Make good flaunching to neighbour's stack with embedded slate drip oversailing rendered face by 20 mm.

9 Plaster: Mix A to exposed brickwork. Plasterboard and Mix C where hearths removed.

9 It may be easier to deal with this area of plaster in the Plaster Repair Schedule.

5.2 New external wall

General

1 It is assumed that large areas of new brickwork will be fully covered by drawings.

2 No finishes are included with new walls.

3 Deep footings to new brickwork should not be bonded to existing work with shallow footings as it may lead to differential settlement problems. Local underpinning may be required.

4 Special clauses to be written wherever tanking, changing wall thicknesses, etc is involved.

5 See L.B. Act Byelaw 7.11 for thickness of walls.

5.2.1 New external solid wall

Location: ....

1 Excavation: For foundation trench 1000 mm deep.

1 The depth of excavation will depend on site soil and will have to be adjusted if necessary unless a trial hole is dug.

2 Concrete: Mix A trenchfill footing 750 mm deep × 600 mm wide.

3 Brickwork: Construct .... brick wall. Facing bricks to match existing adjacent or flettons if to be rendered (see below). Flettons in backings. Solid semi-engineering brick below DPC. English bond. Leadcore DPC 150 mm above ground level lapped with adjacent DPC by 150 mm. Form openings as drawings with PC RC lintels, brick arches and PC concrete cills as scheduled below:

Opening Lintel size Brick arch Concrete cill

3 Insert the thickness of wall.
Insert opening sizes, lintel size and arch and cill details.

4 Brickwork: Second hand facing brick on edge. Coping on two courses tile creasings. Galvanised ms cramps at exposed ends.

Option 4 Brickwork: PC concrete coping bedded on leadcore DPC. 1 no. non-ferrous peg to each section on all raking copings.

4 Select either 4 or option 4.

5 Hardfill: Backfill in hardcore.

6 Render: Mix 1 to window/door reveals externally.

7 Render: Mix 1 to overall externally.

Option 7A Brickwork: Leave raked out ready for pointing to match existing adjacent or new pointing adjacent

Option 7B Brickwork: Flush off mortar bed as work proceeds and hard brush back to show brick edges.

7 Select either 7 or option 7A or 7B.

8 Cross refer: Refer to Drainage Section for lintels etc over drain runs.

9 Cross Refer: Refer to External Wall Section for lateral restraint strap and concrete pads.

5.2.2 New cavity wall

General

Location: ....

This is a model clause, it will require adaptation to suit specific circumstances.

1 Excavation: For foundation trench 1000 mm deep.

2 Concrete: Mix A trenchfill footing 750 mm deep × 600 mm wide.

3 Brickwork: Build wall with 50 mm cavity. Facing brick outer skin to match existing adjacent, 100 mm aerated block inner skin solid semi-engineering brick to both skins below DPC. Stretching bond. Leadcore DPC to each skin 150 mm above ground level (or as shown on drawings) lapped with adjacent injected DPC by 150 mm. Mix A concrete cavity fill to ground level. Weepholes at 900 mm centres. Form openings as drawings with cavity closing PC RC lintels, brick arches and PC concrete cills, on leadcore DPC's as scheduled below, including vertical and horizontal leadcore DPC's as detail drawing no. ....

3 Insert drawing no.
Insert opening sizes and lintel arch and cill details.
Insert galvanised MS or Stainless Steel ties as required by the Local Authority.

Opening	*Lintel size*	*Brick arch*	*Concrete cill*

MS or SS ties at 900 ccs horizontally and 450 vertically throughout cavity work.

4 Brickwork: Construct one brick parapet on leadcore stepped DPC. Facing bricks to match existing adjacent. English bond off one complete course of headers as cavity closer.

5 Brickwork: Brick on edge coping on two courses tile creasings. Galvanised ms cramps at exposed ends.

5 Select either 5, option 5A or option 5B or alternative coping as required.

Option 5A Brickwork: Brick on edge coping on brick oversailing course on leadcore DPC. Galvanised ms cramps at exposed ends.

Option 5B Brickwork: PC concrete coping bedded on leadcore DPC. 1 no. non-ferrous peg to each section on all raking copings.

6 Hardfill: Backfill in hardcore.

7 Render: Mix 1 to window/door reveals externally.

8 Plaster: Mix B overall internally.

Option 8 Plaster: Vermiculite plaster and finish coat internally.

8 Select either 8 or option 8.
Plaster to be specified as required. In this specification plaster on new work is taken with the other work. Only on repaired items is plaster taken in the Plaster Schedule.

9 Pointing: Leave brickwork raked out and point to match existing adjacent or new pointing adjacent.

9 Select either 9 or option 9 or other pointing as required.

Option 9 Pointing: Flush off mortar bed as work proceeds and hard brush back to show brick edges.

10 Cross refer: Refer to Drainage Section for lintels, etc over drain runs.

5.3 Repair existing brickwork

5.3.1 Rebuild defective brickwork (facing bricks)

Location: ....

1 Demolition: Take down defective brickwork. Set aside sound bricks for reuse. Needle and prop brickwork, floors and roof, etc. over to satisfaction of the Local Authority.

2 Brickwork: Rebuild brickwork .... bricks thick. Facing bricks to match existing adjacent. Flettons in backings. Tooth and gauge to existing work. Leave raked out and point in to match existing adjacent or new pointing adjacent.

2 Insert the wall thickness.

3 Brickwork: To include all decorative coursing and corbelling, etc. to match existing and or reconstruct cornice.

4 Brickwork: Form openings as shown to dimensions given or to match existing with PC RC lintels, brick arches and PC concrete cills on leadcore DPC's as scheduled below:

4 Insert opening sizes, lintel sizes and arch and cill details in schedule.

Opening	*Lintel size*	*Brick arch*	*Concrete cill*

5 Brickwork: Brick on edge coping on two courses tile creasings. Galvanised ms cramps at exposed ends.

5 Select either 5 or option 5A or option 5B.

Option 5A Brickwork: Brick on edge coping on brick oversailing course on leadcore DPC. Galvanised ms cramps at exposed ends.

Option 5B Brickwork: PC concrete coping bedded on leadcore DPC with 1 no. non-ferrous peg to each section on all raking copings.

6 Render: Mix 1 to window/door reveals.

7 Carpentry: Cut back joist ends exposed and hang on galvanised ms joist hangers.

8 Cross refer: Refer to External Wall section for lateral restraint straps and concrete pads.

5.3.2 Rebuild defective brickwork (rendered work)

Location: ....

1 Demolition: Take down defective brickwork. Needle and prop brickwork, floors and roof etc over to satisfaction of Local Authority.

2 Brickwork: Rebuild brickwork .... brick thick. Flettons to receive render. Tooth and gauge to existing work.

2 Insert wall thickness.

3 Brickwork: To include all corbelling, etc necessary to reconstruct cornice.

3 Corbelling sometimes contains stone sections. PC concrete or GRP should be considered for very large cornices.

4 Brickwork: Form openings as shown to dimensions given or to match existing, complete with PC RC lintels, brick arches and PC concrete cills on leadcore DPC's as scheduled below:

4 Unless the shape of an arch is required a PC concrete lintel will suffice externally as it will be rendered.
Insert opening sizes, lintel sizes and arch and cill details in schedule.

Opening	*Lintel size*	*Brick arch*	*Concrete cill*

5 Brickwork: PC concrete coping bedded on leadcore DPC with 1 no. non-ferrous peg to each section on all raking copings.

6 Render: Mix 1 overall.

6 Select either 6 or option 6.

Option 6 Render: Mix 1 overall to match original work including all coursing, rustications, cornices, brackets, hoods and the like.

7 Carpentry: Cut back joist ends exposed and hang on galvanised ms joist hangers.

8 Cross Refer: Refer to External Walls section for lateral restraint straps and concrete pads.

5.3.3 Cut out timber wall plates

Location: ....

1 Brickwork: In 1000 mm maximum lengths cut out rotten timber wall plates and piece in fletton brickwork in 1:1:6 mortar.
Allow: .... linear metres one course high of plates carrying joists or rafters.
Allow: .... linear metres one course high of plates within walls.

1 If timbers are sound it may be better to instruct timber treatment subcontractor to carry out precautionary treatment with proprietary paste on timber, or other proprietary means of treating timber in-situ following specialist advice.
Insert the number of linear metres.

5.3.4 Piece in facing bricks

Location: ....

1 Brickwork: Cut out damaged facing bricks and piece in facing bricks to match.
Allow: .... no. bricks.

1 Insert number of bricks.

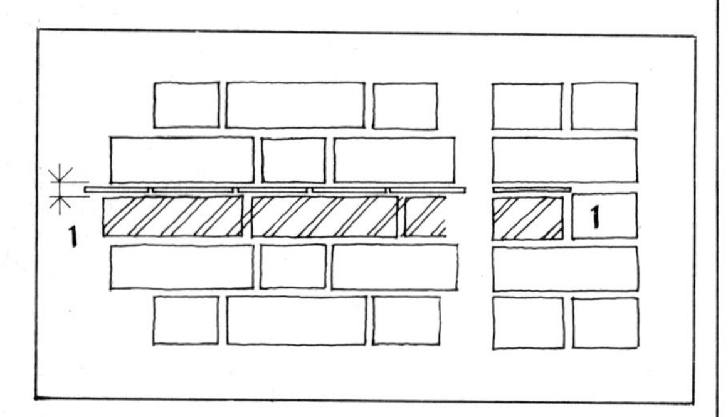

5.3.5 Reinforced concrete corner ties

Location: ....

General

This clause provides for the reinforcement of a failed bond between party walls and party wall or flank walls and front rear walls. It does not replace a lack of bond overall in the wall height. The size of the tie will require adjustment to suit the circumstances.

1 Brickwork: Cut pocket in brickwork to dimensions of tie.

2 Reinforcement: 4 no. 10 mm diameter high yield deformed bars to each tie.

3 Concrete: In-situ Mix B 'L' tie 225 mm high × 112 mm deep extending minimum 450 mm into each wall.

4 Allow: .... ties. Space ties at maximum 1.6 m centres vertically and as shown on drawings.

4 Insert the number of ties to be used. Show the positions of ties on plans and elevations.

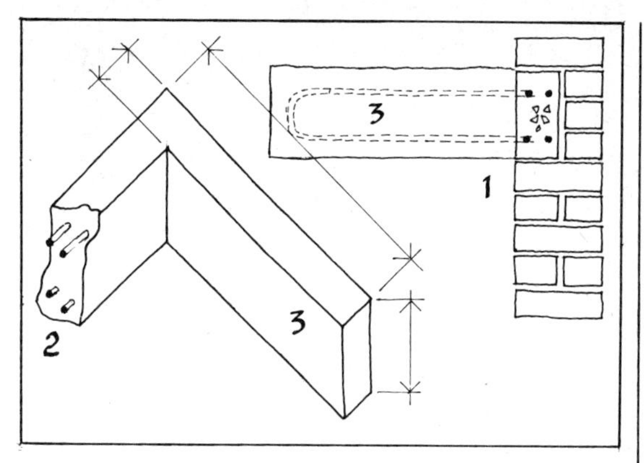

5 This clause is only suitable for use in brickwork with sound mortar and is not suitable for restraining bulges or bows in the brickwork.

5.3.6 CP III lateral restraint straps (direction of joist span)

General

Location: ....

Restraining ties to be used where joist ends are cut and put on galvanised hangers in long runs and in all new construction.
Walls require restraint at both ends—if the other end of joist is on galvanised hangers, straps required both ends to provide wall to wall restraint.

1 Brickwork: Cut out pocket in brickwork to dimensions of pad.

2 Concrete: In-situ Mix B pad surround to strap end. 600 mm long × 150 mm high × 112 mm deep.

2 Pad may not be needed if strap turned down into cavity.

2 & 3 Size strap and fixing to wall to manufacturer's requirements for degree of exposure.

3 Item: Galvanised ms lateral restraint strap size .... (long) .... (turn down) notched in and screwed to joist with 3 no. 12 × 50 mm wood screws. Straps manufactured by ....

....

3 Insert size of strap (usually 850 mm long × 150 mm). Insert manufacturer and note manufacturer's requirements for degree of exposure.
Insert number of ties required.

Allow: .... ties. Space ties one to every third joist, or maximum 1200 mm centres.

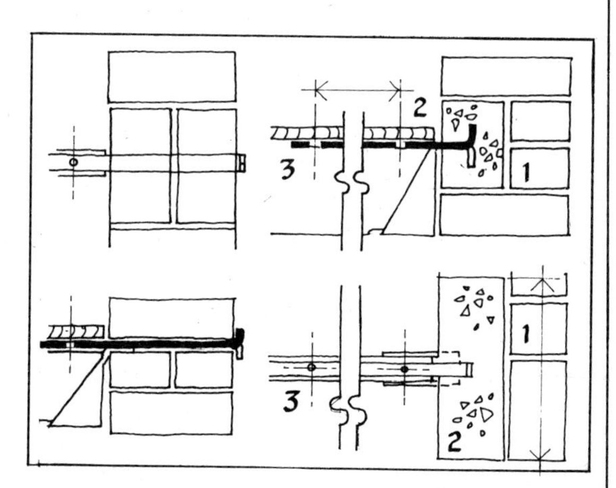

5.3.7 CP III lateral restraint straps (right angles to span)

General

Location: ....

To be used at gable ends on new construction and to restrain old construction showing signs of failure.

1 Brickwork: Cut pocket in brickwork to dimension of pad.

1 In circumstances where brickwork is being rebuilt in the area of the tie end the clause should be adapted to build the tie into the new coursing.

2 Concrete: In-situ Mix B pad surround to strap end. 600 mm long × 150 mm high × 112 mm deep.

3 Item: Galvanised ms lateral restraint strap, size .... mm long × .... mm turn down, notched across 3 no. joists fixed at each joist with 1 no. no. 12 × 50 mm wood screw.
Allow: .... no. ties. Space ties at maximum 1200 mm c/c.

*3 Insert size strap (See **5.3.**7). Note manufacturer's requirements for degree of exposure.*
Insert the number of ties required.

5.3.8 Steel ties

General

Location: ....

This is a general description since the purpose and design of ties is too varied to be selective.

1 Metalwork: Galvanised ms ties as drawing no. .... fixed in position.

1 Insert drawing no.

2 Painting: One coat calcium plumbate primer before fixing.

5.3.9 Expanded metal mesh tie

Location: ....

1 Brickwork: Cut away defective plaster over area. Rake out brick joints to 10 mm depth.

1 Select either 1 or option 1.
This clause is suitable for securing a tie between walls where there are fractures in the bond or continuous lack of bond.
Option 1 For external application.

Option 1 Brickwork: Cut away defective render externally over the area. Rake out brick joints to 10 mm depth.

2 Metal: Apply over the area continuous sheets of galvanised expanded steel. Sheets wired together with galvanised wire at 50 mm ccs along edges, fixed at 225 ccs all over with masonry nails.
Metal specification ....

2 D31 Wrapping fabric to BS 4483/1969 may normally be used. Consult a structural engineer and insert the specification of the metal required.

3 Render: Mix 1 all over area of mesh.

4 Cross Refer: Refer to drawing no. ....

4 Insert drawing number.

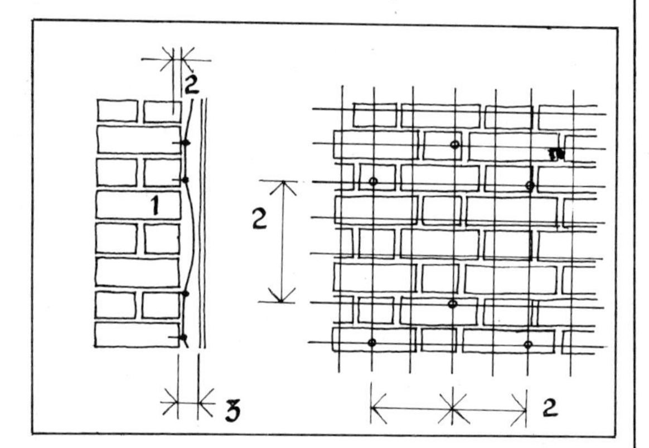

5.3.10 Stitch cracks

Location: ....

1 Demolition: Drill out/chisel out brickwork to one brick depth and 450 mm width along length of crack. Cut back whole and half bricks to follow bonding.

1 Delete drilled or chiselled. Where brickwork is not sound in area of the crack or where adjacent brickwork may be disturbed drilling should be specified.

2 Brickwork: Piece in brickwork following existing bond using similar strength mortar to one existing face of wall.

2 The operation may be repeated both sides of wall in thicknesses greater than 225 mm.

3 Brickwork: Piece in facing brickwork following existing bonding using similar strength mortar and replace pointing to match existing.

4 Allow: .... linear metres.

4 Insert length. Allow for double quantities if both sides of wall to be stitched.

5.4 Blocking openings

5.4.1 Block miscellaneous small openings

Location: ....

1 Demolition: Remove redundant subfloor vents, flues, etc.

2 Brickwork: Block all openings in brickwork to match existing. Point or render to match existing work adjacent.

3 Allow: .... no. openings in one brick wall.

3 Insert the number of minor openings.

4 Allow: .... no. openings in one and a half brick wall.

4 Insert the number of minor openings.

5.4.2 Block fireplace

General

Location: ....

This clause assumes the removal of fireplace has been covered in Demolitions clause. Plaster is assumed to be included elsewhere. 2 sm of plaster should be allowed for each blocked fireplace.

1 Item: Sweep flue.

1 If there are a number of flues being blocked then the sweeping of flues may be gathered in one clause.

2 Brickwork: Block opening in half brick wall.

3 Item: Lead cored DPC lapped 150 mm with adjacent DPC (at lowest floor level).

3 This applies to the blocking of flues at ground level and below.

5.4.3 Block fireplace (for gas fire)

General

Location: ....

See notes above.

1 Item: Sweep flue.

2 Brickwork: Block opening in half brick wall. Leave opening for gas fire closer plate, minimum opening 300 mm wide × 500 mm high × 165 mm deep.

3 Item: Lead cored DPC lapped 150 mm with adjacent DPC (at lowest floor level).

5.4.4 Block fireplace (for back boiler)

General

Location: ....

See notes above.

1 Item: Sweep flue

2 Brickwork: Fletton or block plinth to boiler manufacturer's requirements with openings for pipework.

3 Brickwork: Block opening in half brick wall (on leadcore DPC lapped 150 mm with adjacent DPC at lowest floor level). Leave opening for back boiler to manufacturer's requirements.

4 Concrete: After installation of flue liner, cast concrete around liner to seal flue above boiler.

4 The flue sealer ensures that the boiler is not damaged by soot or debris falling down the flue and that the liner is fixed to the boiler.

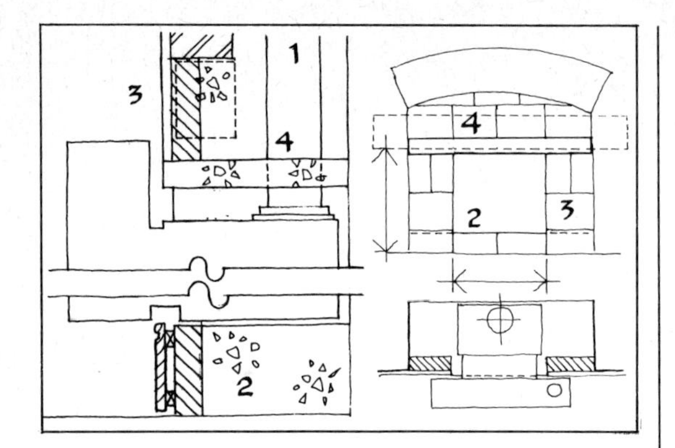

5.4.5 Block flues (fireplace as recess)

Location: ....

1 Item: Sweep flue.

2 Carpentry: 75 × 50 mm tanalised sw noggings to follow line of arch with galvanised expanded metal lath fixed to soffit.

3 Plaster: Mix A to metal lath and recess.

3 At ground level and below consider DPC or tanking details. Salt resistant render may be required to stop salts from soot coming through.

5.4.6 Block external openings (with new footings)

Location: ....

1 Demolition: Take out stone cill, brick arch and lintel and brickwork under opening. Prop over.

1 This clause extends and ties into the existing footings either with concrete or by extending the existing footing.

2 Excavations: For foundation trench to level of existing footings adjacent or to Local Authority approval.

3 Concrete: Mix B footing in trench 300 mm deep × 600 mm wide.

3 Select either 3 or option 3.

Option 3 Brickwork: 50 mm Mix A concrete blinding to bottom of trench. Semi-engineering spread brick footing to match existing adjacent, bonded to existing. Heading bond stepping 28 mm per course.

Option 3 Applies in houses below 4 storeys subject to the approval of Local Authority Inspector. Extending the existing footing can only be done where these are on firm clay, and not subject to sulphate attack.

4 Brickwork: Block opening .... bricks thick. Facing bricks to match existing adjacent or flettons if to be rendered (see below). Flettons in backings. Semi-engineering bricks below ground. Tooth and gauge to existing work. Lead core DPC at level of adjacent DPC lapped by 150 mm.

4 Insert the wall thickness.

5 Hardfill: Backfill in hard core.

6 Brickwork: Leave raked out and point in to match existing adjacent or new pointing adjacent.

6 Select either 6 or option 6.

Option 6 Render: Mix 1 overall to match adjacent work including all coursing and rustications.

7 Cross refer: Refer to Drainage Section for lintels, etc. over drain runs.

5.4.7 Block external opening (on existing footings or at high level).

Location: ....

1 Demolition: Take out stone cill, brick arch and lintel. Prop over.

2 Brickwork: Block opening .... bricks thick. Facing bricks to match existing adjacent or flettons if to be rendered (see below). Flettons in backings. Gauge to existing work. When opening is adjacent to ground level provide leadcore DPC at level of DPC adjacent, lapped by 150 mm.

2 Insert the wall thickness.

3 Brickwork: Leave raked out and point to match existing adjacent or new pointing adjacent.

3 Select either 3 or option 3.

Option 3 Render: Mix 1 overall to match adjacent work including all coursing and rustications.

5.5 Forming/altering openings

5.5.1 Form new opening/adjust existing opening

Location: ....

1 Demolition: Cut opening in brickwork and prop over.

2 Brickwork: Quoin up jambs in facing bricks to match existing adjacent or flettons if to be rendered (see below). Tooth and gauge to existing work. PC RC lintel, brick arch and PC concrete cill on leadcore DPC as scheduled below.

2 This clause covers facing brick walls. Insert the opening sizes, lintel sizes and arch and cill details.

Opening Lintel size Brick arch Concrete cill

3 Brickwork: Leave raked out and point in to match existing adjacent or new pointing adjacent.

3 Select either 3 or option 3.

Option 3 Render: Mix 1 into opening to match adjacent work including all coursing and rustications.

4 Render: Mix 1 to jambs and arch soffit.

4 For render to reveals only.

5.5.2 Raise cill height (with new footings)

Location: ....

1 Demolition: Take out existing stone cill and brickwork under opening.

2 Excavations: For foundation trench to level of existing footings adjacent, or to Local Authority approval.

3 Concrete: Mix B footing in trench 300 mm deep × 600 mm wide.

3 Select either 3 or option 3.

Option 3 Brickwork: 50 mm Mix A concrete blinding to bottom of trench. Semi-engineering spread brick footing to match existing adjacent, bonded to existing. Heading bond to match existing.

4 Brickwork: Partly block opening .... bricks thick. Facing bricks to match existing adjacent. Flettons in backings. Semi-engineering bricks below ground. Tooth and gauge to existing work. Leadcore DPC at level of adjacent DPC, lapped with adjacent DPC by 150 mm. PC concrete cill on leadcore DPC. New cill height .... mm above FFL. Rake out and point to match existing adjacent or new pointing adjacent.

4 Select either 4 or option 4.
Insert the wall thickness.
Insert the new cill height.
This clause is for use with brick faced walls.

Option 4 Brickwork: Partly block opening .... bricks thick. Flettons to receive render. Semi-engineering bricks below ground. Tooth and

Option 4 Insert the wall thickness.
Insert the new cill height.
This clause is for use with render faced walls.

gauge to existing work. Leadcore DPC at level of injected DPC, lapped with injected DPC by 150 mm. PC concrete cill on leadcore DPC. New cill height .... mm above FFL.

5 Hardfill: Backfill in hardcore.

6 Render: Make good rendered reveals.

7 Render: Mix 1 overall to match adjacent work including all coursing and rustication.

7 For use with option 4 only.

8 Cross refer: Refer to Drainage Section for lintels etc over drain runs.

5.5.3 Raise cill height (on existing footings or at high level)

Location: ....

1 Demolition: Take out existing stone cill.

2 Brickwork: Partly block opening .... bricks thick. Facing bricks to match existing adjacent. Flettons in backings. Tooth and gauge to existing work. (When opening is adjacent to ground level provide leadcore DPC at level of injected DPC adjacent, lapped with injected DPC by 150 mm). PC concrete cill on leadcore DPC. New cill height .... mm. Rake out and point to match existing adjacent or new pointing adjacent.

2 Select either 2 or option 2. Insert the wall thickness. Insert the new cill height. This clause for use with brick faced walls.

Option 2 Brickwork: Partly block opening .... bricks thick. Flettons to receive render. Tooth and gauge to existing work. (When opening is adjacent to ground level provide leadcore DPC at level of injected DPC adjacent, lapped with injected DPC by 150 mm.) PC concrete cill on leadcore DPC. New cill height .... mm above FFL.

Option 2. Insert the wall thickness.
Insert the new cill height.
This clause for use with render faced walls.

3 Render: Make good rendered reveals.

4 Render: Mix 1 overall to match adjacent work including all coursing and rustication.

4 For use with option 2 only.

5.5.4 Lower cill height

Location: ....

1 Demolition: Take out existing stone cill and brickwork below to extent shown.

2 Brickwork: PC concrete cill on leadcore DPC. Make good brickwork at reveals to match existing. New cill height .... mm above FFL.

2 Insert cill height.

3 Render: Make good in Mix 1 to rendered reveals.

4 Render: Make good in Mix 1 to adjacent work including all coursing and rustication.

4 For use with rendered walls only.

5.5.5 Replace existing lintels/brick arches

Location: ....

1 Demolition: Take out brick arch and lintel and prop over.

2 Brickwork: PC RC lintel and brick arch as scheduled below:

2 Insert the opening size, lintel size and arch detail.

Opening	*Lintel size*	*Brick arch*

3 Render: Make good in Mix 1 to rendered reveals.

4 Render: Mix 1 to match original work over opening.

4 For use with rendered walls and/or reveals only.

5.5.6 Replace concrete cill

Location: ....

1 Demolition: Take out existing stone cill.

2 Brickwork: PC concrete cill laid on leadcore DPC size .... Make good all brickwork and render disturbed to match existing. Check the lintel size on site.

2 Insert cill size.

3 Cross refer: Refer to drawing no. ....

3 Insert drawing no.

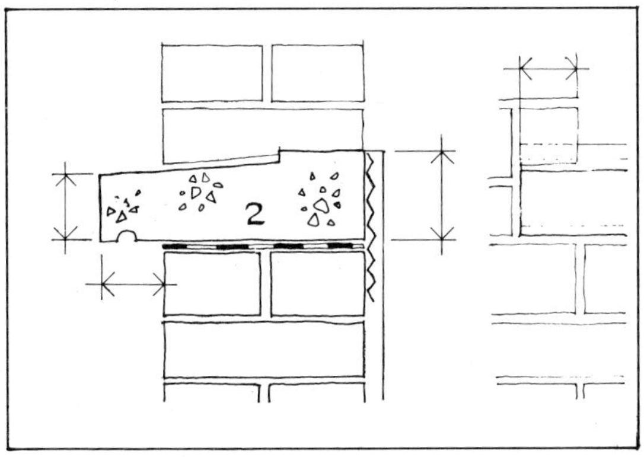

5.5.7 Replace concrete cill (reinforced)

Location: ....

1 Demolition: Take out existing stone cill.

2 Concrete: PC concrete cill reinforced with steel mesh to BS 4483 Code C 283 cut to provide two longitudinal bars, 100 × 400 mm laid on lead core DPC, section to match existing but with 12 mm half round drip under. Minimum thickness of cill to be 90 mm.

2 British Standards recommend that all cills over 600 mm length should be reinforced for handling. It is advisable that reinforcement is galvanised.

3 Cross refer: Refer to drawing no. ....

3 Insert drawing no.

5.5.8 Brick up recessed reveals

Location: ....

General

This clause is used where a casement window is to replace a sash box or where shutter casings have been removed and are to be blocked up.

1 Brickwork: Brick up existing recessed reveals in flettons bonded to existing work including toothing in to external skin with facing bricks every 600 mm vertically. Make good all rendered work disturbed.

5.5.9 Airbrick/airvent (225 × 150)

Location: ....

General

This clause describes the typical airbrick required by the London Building and Public Health Acts for habitable rooms combined with Clause **5.5.**10, *where a boiler or room appliance requires larger areas of ventilation.*

1 Brickwork: Form opening and fit 225 × 150 mm terracotta/galvanised CI airvent externally. Make good brickwork/render. Fit 225 × 150 mm plaster louvre vent internally. Opening to be not less than 1.75 mm above floor level.

1 Insert terracotta or galvanised CI as required.

5.5.10 Airbrick/airvent

Location: ....

1 Brickwork: Form opening and fit 225 × .... mm terracotta airbrick on external face of wall. Make good brickwork/render. Internally fit 225 × .... mm plaster louvre ventilator. Openings for permanent room ventilation to be at high level.

1 Insert louvre size, either 150 or 225 mm. This gives free air area of 9742 sq. mm. which is adequate if boiler not at full output. Check on local requirements. Typical Gas Board regulations for venting rooms containing boilers are as follows:

Position of Vent	*Conventional Flue*		*Balanced Flue*	
	Air From Room	Air From Outside	Air From Room	Air From Outside
All at high level	900mm²/ Kw 2in²/ 5000 Btu	450mm² Kw 1in²/ 5000 Btu	900mm²/ Kw 2in²/ 5000 Btu	450mm² Kw 1in²/ 5000 Btu
OR All at low level	1800mm²/ Kw 4in²/ 5000 Btu	900mm² Kw 2in²/ 5000 Btu	900mm²/ Kw 2in²/ 5000 Btu	450mm² Kw 1in²/ 5000 Btu

All figures for free area of vent.

Take care in the location of airbricks in loadbearing piers, particularly where free area of venting needed requires more than one airbrick opening.

Typical Free Area for BS 493: Part 2.

Airbricks

225 × 75	*1290 mm²*
225 × 150	*2580 mm²*
225 × 225	*4500 mm²*

Other alternatives are CI airbricks or slot ventilators in window frames.

5.5.11 Airbrick/airbrick

Location: ....

1 Brickwork: Form opening and fit 2 no. 225 mm × .... terracotta airbricks, 1 no. on external face, 1 no. on internal face of wall. Make good brickwork/render.

1 Insert size 75, 150 or 250 mm.
For venting under steps and meter cupboards etc. check that the free area of the airbrick will give a venting area which will comply with local requirements.

5.5.12 Airvent in chimney breast

General

Location: ....

For use when fireplace has been blocked up.

1 Brickwork: Locate flue serving room to be vented by approved means. Form opening at high level and fit 225 × 75 mm plaster louvre vent.

1 Check that the level of the flue vent meets local requirements.

5.5.13 Balanced flue boiler outlet

General

Location: ....

Check that the location of the outlet is acceptable to the Gas Board.

1 Brickwork: Form opening to suit balanced flue boiler outlet with embedded slate over to support brickwork. Make good brickwork/render. Clear opening size .... mm high × .... mm wide.

Select either 1 or option 1.
1 Insert the specified boiler flue opening size.

Option 1 Brickwork: Insert PC concrete lintels size .... to opening .... × .... for flue.

Option 1 This clause serves where the balanced flue is larger than standard opening and requires lintel support to opening. Insert the lintel size and balanced flue opening required below it.

2 Item: Flue to be supplied and fitted by central heating installer.

5.6 Parapets and stacks

5.6.1 Repair parapets

Location: ....

1 Demolition: Take down defective brickwork to parapets. Set aside sound bricks for reuse.

2 Brickwork: Build up parapets one brick thick. Facing bricks to match existing adjacent toothed and gauged to existing work including all corbelling and the like and all raking cuts.

3 Cross refer: Refer to Clause for copings.

4 Allow: .... m^2

4 Insert the area to be rebuilt.

5.6.2 Copings (brick on creasing tiles)

Location: ....

1 Brickwork: Take off existing copings. Lay 2 no. courses clay creasing tiles and facing brick on edge coping to match existing adjacent. Exposed ends cramped with galvanised ms cramps.

2 Allow: .... linear metres.

2 Insert the number of linear metres.

5.6.3 Copings (brick on sailing course)

Location: ....

1 Brickwork: Take off existing copings and sailing course. Lay facing brick to match existing adjacent on edge coping on brick oversailing course on leadcore DPC. Galvanised ms cramps at exposed ends.

2 Allow: .... linear metres.

2 Insert the number of linear metres.

5.6.4 Copings (concrete)

Location: ....

1 Brickwork: Take off existing copings. Lay .... weathered PC concrete coping with half round drips on leadcore DPC. 1 no. non ferrous peg to each section on all raking copings.

1 Insert 'once' or 'twice' weathered.

2 Allow: .... linear metres.

2 Insert the number of linear metres.

5.6.5 Copings (tile)

Location: ....

1 Brickwork: Re-bed all loose coping tiles and replace all those that are broken. Replacement tiles to match existing. Rake out and point all joints between soundly fixed tiles.

1 Many houses have tiled copings which can be taken off and re-bedded. Alter the clause if necessary.

2 Allow: .... no. replacement tiles.

2 Insert number of new tiles to be allowed for.

5.6.6 Rebuild stack (retaining pots)

General

Location: ....

If stack is shared the work is subject to agreement with the adjoining owner.

1 Demolition: Set aside sound existing pots. Take off flaunching. Take down .... no. courses brickwork.

1 Insert number of courses to be taken down.

2 Brickwork: Rebuild in facing bricks to match existing including all sailing courses. Reduce flues as necessary with slates. Re-bed and flaunch pots, set aside. Allow .... no new pots to match existing. Locate, by approved means, flues that served house fireplaces now blocked up, and fit loose spigotted birdproof terminals (........ no) into pots (flue ventilator type).

2 Insert the number of new pots and size and number of new terminals.

3 Brickwork: Fit .... no. spigotted terminals to pots serving boiler flues in accordance with manufacturer's directions.

Manufacturer ....

Cat no. ....

3 This allows the existing pots to be retained and used in conjunction with boiler flues.
Insert the number of new terminals and manufacturer and number.
Terminals must be approved by Gas Council.
Terracotta terminals are available.

5.6.7 Rebuild stack (removing pots)

General

Location: ....

If the stack is shared the work is subject to agreement with the adjoining owner. Consider whether or not to keep the existing chimney pots if sound and insert spiggots direct to existing pots.

1 Demolition: Take off flaunching and pots, setting aside sound pots for use on neighbour's side. Take down .... no. courses brickwork.

1 Select either 1 or option 1.
Insert number of courses to be taken down.

Option 1 Demolition: Take off flaunching and pots. Take down .... no. courses brickwork.

Option 1. Insert the number of courses to be removed.

2 Brickwork: Rebuild in facing bricks to match existing, including all sailing courses. Reduce flues as necessary with slates. Re-bed and flaunch means, flues that served house fireplaces now blocked up, and bed and flaunch .... no. new birdproof terminals. Re-bed and flaunch neighbour's pots set aside.

2 Select either 2 or option 2.
Insert the number of terminals to be fixed.

Option 2 Brickwork: Rebuild in facing bricks to match existing, including all sailing courses. Reduce flues as necessary with slates. Bed and flaunch .... no. new birdproof terminals.

Option 2 This option applies to stacks not affecting neighbours' flues.
Insert the number of birdproof terminals.

3 Brickwork: Bed and flaunch .... no. terminals for boiler flues in accordance with manufacturer's directions.

3 Insert the number and specification of boiler flue terminals.

Manufacturer ....

Cat. no. ....

5.6.8 Take down stack and cap off

General

Location: ....

If the stack is shared the work is subject to agreement with adjoining owner.

1 Demolition: Take off pots and take down brickwork to parapet level.

2 Brickwork: Form openings in side of stack to each flue and fit 225 × 225 mm terracotta louvred airbrick (........ no). Cap off stack with 50 mm thick PC concrete paving slabs bedded to falls, laid on leadcore DPC. Slabs to oversail brickwork by 50 mm all round.

2 Louvred airbricks are required in exposed positions to stop rain penetration. Insert number.

5.6.9 Cap off existing pots

Location: ....

1 Brickwork: Repair existing flaunching. Locate by approved means, flues that served house fireplaces now blocked up and fit .... no. loose spigotted birdproof terminals in existing pots (flue ventilator type).

1 Insert the number of spigotted terminals required.

2 Brickwork: Fit .... no. spigotted terminals to pots serving boiler flues in accordance with manufacturer's directions.

2 Insert the number of boiler flue terminals required.
Insert type and manufacturer.
Terminal to be to the Gas Council or Gas Board approval.
Terracotta terminals are available.

Manufacturer ....

Cat. no. ....

5.6.10 Remove existing pots

Location: ....

1 Brickwork: Locate by approved means, flues that served house fireplaces now blocked up. Carfully remove existing pots. Bed and flaunch .... no. birdproof terminals.

1 Insert the number of birdproof terminals required.

2 Brickwork: Bed and flaunch .... no. terminals for boiler flues in accordance with manufacturer's directions.

2 Insert thc number of boiler flue terminals required. Insert type and manufacturer.

Manufacturer. ....

Cat. no. ....

5.7 External wall finishes

5.7.1 Pointing

Location: ....

1 Brickwork: Rake out existing coursing to 20 mm depth and repoint brickwork at any level as directed on site in .... pointing.

1 Insert the type of pointing required. The most common is weatherstruck.

2 Allow: .... m²

2 Insert the area of pointing to be done.

5.7.2 Rendering (replacing existing render)

General

Location: ....

To avoid existing render bridging an injected DPC.

1 Render: Hack off defective render.

2 Render: Rake out brickwork over area to be rendered. Render Mix .... Stop off 150 mm above ground level on galvanised ms bell mouth render stop bead.

2 Insert the render mix required.
Mix 1 is plain
Mix 2 is pea gravel finish.

3 Brickwork: Rake out and point brickwork below stop and paint exposed plinth 2 no. coats bituminous paint blinded with sharp sand.

4 Allowance .... m² hacking off and .... m² new render.

4 Insert quantities.

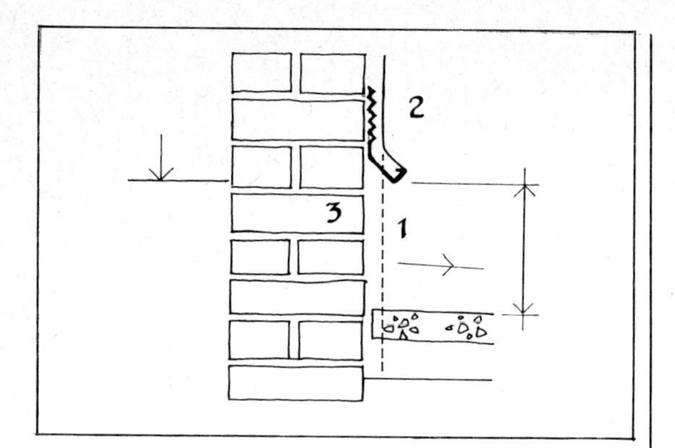

5.7.3 Rendering (to existing facing brick)

Location: ....

1 Render: Rake out existing brickwork over area to be rendered. Render in Mix 1 finish. Stop off 150 mm above ground level on galvanised ms bell mouth render stop bead.

1 Delete the finish not required. Insert either woodfloat or steel trowelled finish.

2 Render: Rusticated and coursed work to be set out as existing or as drawing no. ....

2 Not to be used if plain render only. Insert drawing number if appropriate.

3 Brickwork: Rake out and repoint below stop bead and paint plinth 2 coats bituminous paint blinded with sharp sand.

4 Allowance .... m^2 hacking off and .... m^2 new render.

4 Insert quantities.

6 Roofs

6.1 New flat roof structure
6.1.1 Flat roof structure for built-up roofing, asphalt, lead or zinc finish

6.2 New pitched roof structures
6.2.1 New pitched roof structure

6.3 Stripping roof coverings/repairs to existing roof cladding and structures
6.3.1 Strip roof finish/repair decking
6.3.2 Strip slate or tile roof/repair timbers

6.4 Hatches/eaves/soffits
6.4.1 Roof access hatch: pitched roof
6.4.2 Eaves boards
6.4.3 Soffit boards

6.5 Asphalt roofing/paving (including associated leadwork)
6.5.1 Asphalt roof finish to timber deck
6.5.2 Asphalt roof finish to concrete deck
6.5.3 Asphalt finish to external steps/landings

6.6 Built-up roofing (including associated leadwork)
6.6.1 Built-up roof finish

6.7 Zinc and lead sheet roofing
6.7.1 Zinc roof finish
6.7.2 Lead roof finish

6.8 Natural/asbestos cement slates and concrete tiles including associated zinc and leadwork
6.8.1 Natural slate roof
6.8.2 Asbestos cement/slate roof
6.8.3 Concrete tile roof
6.8.4 Lead/zinc flashings and soakers etc. for slate/tiled roof

6.9 Repairs to existing roof coverings
6.9.1 Repairs to existing flat roof covering
6.9.2 Repairs to existing pitched roof covering
6.9.3 Relay valley gutter

6.10 Insulation and protective finishes
6.10.1 Flat roof insulation for built-up roofing or asphalt finish (warm roof)
6.10.2 Flat roof insulation for built-up roofing or asphalt finish ('upside down' roof)
6.10.3 Flat roof insulation for lead or zinc roof finish ('cold' roof)
6.10.4 Pitched roof insulation
6.10.5 Concrete paving slabs on slab supports for trafficable 'upside down' roof
6.10.6 Gravel finish for non-trafficable 'upside down' roof
6.10.7 Reflective stone chippings for non-trafficable built-up roofing or asphalt finish
6.10.8 Reflective paint for non-trafficable built-up roofing or asphalt finish

6 Roofs

6.1 New flat roof structure

6.1.1 Flat roof structure for built-up roofing asphalt, lead or zinc finish

Location: . . . .

General

Roof coverings and insulation are described elsewhere.
NOTE: Insulation and roof finish must all be laid on the same day. Insulation must be kept dry.

1 Carpentry: . . . . mm × 50 mm tanalised sw joists at 400 mm centres hung on galvanised joist hangers to brickwork.

1 Insert the joist depth.

2 Carpentry: Tanalised sw firring pieces to same width as joists spiked to joists to give minimum fall of 1 in 40.
Minimum thickness of 25 mm.

2 Firring pieces are not necessary if the roof is to have an 'upside-down' finish (see Clause ***6.10.****2)*

3 Carpentry: 38 × 38 mm tanalised sw herringbone strutting at . . . . mm centres.

3 Insert spacing of strutting. Strutting should be used where spans exceed 50 times the joist width. A single row should be used for over 2.5 m span and two rows for spans over 5 m and up to 7.5 m.

4 Carpentry: 19 mm exterior WBP ply nailed at 150 mm centres to all edges and intermediate supports, with 50 × 50 mm tanalised sw noggins under all cross joints. Leave 6 mm clearance at perimeter.

Option 4 Carpentry: Ex 25 mm × 100 mm tanalised sw board cramped and twice nailed to each joist with galvanised nails. Stagger joints, leave 6 mm clearance at perimeter.

4 Select either 4 or option 4.
Boards should be used in the maximum sizes possible to minimise the frequency of joints.

5 Carpentry: Form kerb all round perimeter of deck of ex 175 × 50 mm splayed tanalised sw kerb securely nailed with galvanised nails to joists and noggings. Leave 6 mm clearance between rear face and brickwork. Do not fix to brickwork.

5 A free standing kerb is essential in all cases to permit the roof covering to be isolated from the main structure allowing differential movement of deck and structure. A cover flashing specified elsewhere protects the top. The kerb should be omitted from any sides not having a brick upstand wall.

6 Carpentry: Form opening in boarding for proprietary rainwater outlet all as shown on drawing no. . . . .

6 Select either 6 or option 6.
Insert drawing number or delete.

Option 6 Carpentry: Form catchpit and chute as shown on drawing no. . . . . to discharge rainwater into hopper.

Option 6 Insert drawing number or delete.

7 Carpentry: Drill 25 mm holes at 400 mm centres at mid depth of all joists.

7 This is only necessary for a 'cold' roof. See General note for clauses ***6.7.****1 and* ***6.7.****2.*

It is also necessary to ventilate the roof void to the outside air. This may be done with proprietary perpend vents. This ventilation should be specified in the External Wall Section.

8 Plasterboard: 9.5 × 1200 × 406 mm foil backed Gyproc lath to underside of joists including 38 × 38 mm tanalised noggings at edges.

8 Thistle base board is an alternative to Gyproc lath. Various sizes are available in each, but it is important to relate board length to joist spacing. Plaster board should be omitted if a sound reducing structurally separated ceiling is being installed if the ceiling joists are staggered with the roof joists. If a sound reducing ceiling is below the level of the joists the inclusion of plasterboard to the underside of the roof joists will increase the sound reduction.

9 Plaster: Jute scrim to internal angles. Thin-coat metal angle bead to external angles. 5 mm Thistle board finish coat, smooth finish.

9 Omit if plasterboard is omitted or if it is used as a lining for sound reduction above a separate ceiling structure.

10 Cross refer: Refer to External Walls section for restraining straps and this section for specification of insulation and roof finish.

11 Cross refer: Refer to drawing no. ....

11 Insert drawing no.

6.2 New pitched roof structure

6.2.1 New pitched roof structure

Location: ....

1

1 This Clause should be specifically written for each case.

6.3 Stripping roof coverings/repairs to existing roof cladding and structures

6.3.1 Strip roof finish/repair decking

General

Location: ....

This clause is for flat roofs and bay window roofs.

1 Demolition: Strip existing roof covering including all upstands and cover flashings.

2 Carpentry: Replace defective joists, rafters and plates in 100 × 50 mm tanalised sw.
Allow .... linear metres.

2 Insert allowance required.

3 Carpentry: Replace defective joists in 150 × 50 mm tanalised sw.
Allow .... linear metres.

3 Insert allowance required.

4 Carpentry: Replace defective joists in 200 × 50 mm tanalised sw timber.
Allow .... linear metres.

4 Insert allowance required.

5 Carpentry: Replace defective boarding and cladding in 150 × 25 mm tanalised sw plain edge boarding.
Allow .... linear metres.

5 Insert allowance required.

6.3.2 Strip slate or tile roof/repair timbers

Location: ....

1 Demolition: Strip existing slates or tiles setting aside sound slates/tiles for reuse. Strip off all battens, cement fillets, flashings, soakers, valley gutter linings, etc.

1 Select 1 or option 1.
Select 1 when the roof is to be re-roofed with second hand slates or tiles of the same size as existing.

Option 1 Demolition: Strip existing slates or tiles. Allow in pricing for resale value of sound slates or tiles. Strip off all battens, cement fillets, flashings, soakers, valley gutter linings, etc.

2 Demolition: Carefully take off all ridge tiles and set aside for re-use.

2 Include clause 2 only if the ridge tiles are worth saving. Specify their re-use in clause **6.8.**1 *or* **6.8.**2.

3 Demolition: Carefully take off all hip tiles and set aside for re-use.

3 Include clause 3 only if the hip tiles are worth saving. Specify their re-use in clause **6.8.**1 *or* **6.8.**2.

4 Carpentry: Replace defective joists, rafters and plates in 100 × 50 mm tanalised sw.
Allow .... linear metres.

4 Insert allowance required.

5 Carpentry: Replace defective joists in 150 × 50 mm tanalised sw.
Allow .... linear metres.

5 Insert allowance required.

6 Carpentry: Replace defective joists in 200 × 50 mm tanalised sw.
Allow: .... linear metres.

6 Insert allowance required.

7 Carpentry: Replace defective boarding and cladding in 150 × 25 mm tanalised plain edge boarding.
Allow: .... linear metres.

7 Insert allowance required.

6.4 Hatches/eaves/soffits

6.4.1 Roof access hatch: Pitched roof

Location: ....

1 Demolition: Strip out existing lining and hatch.

1 Select either 1 or option 1.
Use 1, when there is an existing hatch.

Option 1 Carpentry: Trim opening between roof members in tanalised sw to same dimensions as adjacent members.

2 Carpentry: Form 12 mm WBP ply kerb with ex 25 × 50 mm tanalised sw frame screwed to frame and around upper edge. 12 mm WBP external ply hatch with 25 × 50 mm tanalised sw downstand frame around perimeter to master kerb. 2 no. 150 mm steel barrel bolts at opposite corners internally. 2 no. 150 mm steel grab handles.

2 Specify weather-proofing finish in zinc or lead sheet in the Roofing Section.

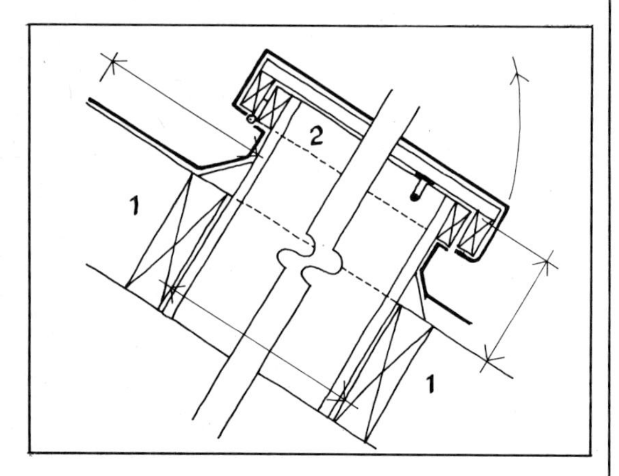

6.4.2 Eaves Boards

Location: ....

1 Demolition: Strip off existing eaves boards.

2 Carpentry: Plug and screw to brickwork using plastic plugs and non-ferrous screws, ex 175 × 25 mm sw eaves boarding to master soffit board (if any) by 12 mm. Mitred corners. Include for trimming rafter ends and fixing tanalised backings as necessary to give a true line. Prime before fixing.

3 Carpentry: 12 × 25 mm tanalised sw packing pieces to eaves boards at minimum 500 mm centres plugged and screwed to brickwork using plastic plugs and non-ferrous screws.

3 Use Clause 3 when there is no soffit to allow air to circulate around rafter ends, wallplate and also to vent roof space. Insulation should not be allowed to block ventilation.

4 Painting: Prime overall before fixing.

6.4.3 Soffit boards

Location: ....

1 Demolition: Strip off existing soffit boards.

2 Carpentry: 9 mm exterior WBP ply to soffit including tanalised bearer to give a true line.

2 Select either 2 or option 2.

Option 2 Carpentry: Ex 19 × 100 mm tongued and grooved sw boarding to soffit with mitred corners. Included tanalised bearers to give a true line. Prime before fixing.

3 Carpentry: Ensure that there is a gap of minimum 12 mm and maximum 19 mm between rear line of soffit and face of brickwork.

3 The gap allows air to circulate as with Option 3 of ***6.4.****2 above.*

4 Painting: Prime overall before fixing.

6.5 Asphalt roofing/paving (including associated leadwork)

6.5.1 Asphalt roof finish to timber deck

Location: ....

General

It is preferable to avoid exposed edges if possible using upstands all round. With upstands it is important to ensure that rainwater outlets are adequate and effective. Before writing consider the insulation.

1 Item: Thoroughly clean area to be asphalted.

2 Brickwork: Rake out brickwork joint immediately above (but not less than 12 mm above) highest point of top of timber kerb to receive lead cover flashing. Depth 25 mm.

2 Select either 2 or option 2.

Option 2 Render: 25 mm above highest point of top of timber kerb cut, chase, using a disc cutter to receive lead cover flashing. Depth 25 mm.

Option 2 Select Option 2 if the upstands have sound render.

3 Felt: Lay isolating membrane to BS 747. Type 4A(i) black bitumen sheathing felt, weight 17 kg per standard 25 m roll, laid loose with 50 mm laps including over kerbs (if any). Fix galvanised expanded metal with galvanised nails to full height of kerbs and 100 mm horizontally across roof deck.

4 Lead: Fix 150 mm girth Code 5 lead flashing along eaves at 250 mm centres with galvanised clout nails to 75 mm × (thickness of insulation, less 12 mm) tanalised plate securely nailed to roof deck. Flashing to be taken 75 mm across plate and dressed into gutter. Welt both edges and fix gutter edge with lead clips at 600 mm centres. Prime with bitumen primer before asphalt laid, leaving exposed leadwork clean.

4 Include Clause 4 if the roof discharges straight into a gutter and not via a rainwater outlet.

5 Asphalt: Lay 20 mm roofing asphalt to BS 988 in 2 equal layers with minimum 150 mm laps. Dress up over kerbs with 2 coat angle fillet.

5 See Appendix.

6 Asphalt: Form fair bullnosed edge at all eaves, minimum 32 mm thick.

6 Include Clause 6 unless there is an upstand all round.

7 Lead: Dress Code 4 lead cover flashing over all kerbs. Girth to be 150 mm. Tuck into chase 25 mm. Wedge, point and clip with 50 mm wide Code 4 lead tacks at minimum 600 mm centres. Maximum length any piece 2.1 m with ends lapped 100 mm. Vertical stepped flashing at exposed ends.

7 Other less expensive materials can be used in place of lead.

8 Cross refer: Refer to detail drawing no ....

8 Insert drawing no.
Detailed drawings are advisable.

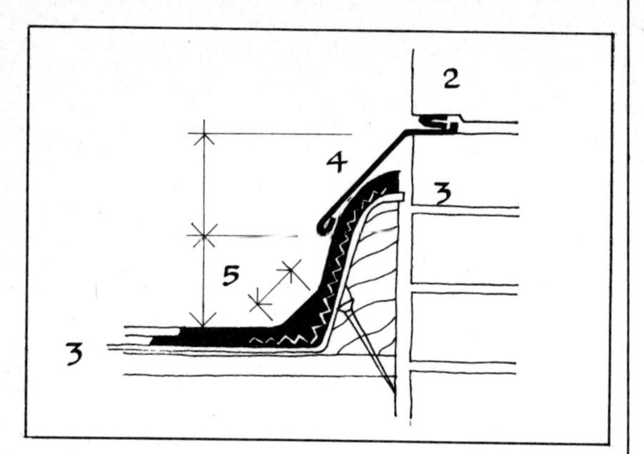

6.5.2 Asphalt roof finish to concrete deck

General

Location: ....

The General note to ***6.5.****1 applies.*

1 Item: Thoroughly clean area to be asphalted.

2 Brickwork: Form 25 × 25 mm splayed chase to all brick upstands with top of chase 150 mm above highest point of concrete roof slab or insulation.

2 Select either 2 or option 2.
This assumes the existence of sound brickwork. It may be necessary to repair it prior to asphalting.

Option 2 Brickwork: Hack off render to all rendered upstands up to an accurately cut line 150 mm above highest point of deck or insulation. Dub out exposed brickwork on 1:1:6 render with scribed surface to receive asphalt. Form 25 × 25 mm splayed chase with top of chase at line cut in render.

Option 2 Select Option 2 if any upstands have render sound enough to retain above the line of the asphalt upstands. If the render is not sound enough the work to the render should be included in the clause specifying render repairs.

3 Felt: Lay isolating membrane to BS 747, Type 4A (i) black bitumen sheathing felt, weight 17 kg per standard 25 m roll, laid loose with 50 mm laps.

4 Asphalt: Lay 20 mm roofing asphalt to BS 988 in 2 equal layers with minimum 150 mm laps. Form 150 mm high upstand(s) to brickwork 20 mm thick with 2 coat angle fillets. Tuck into chases. Point top of chase on completion.

5 Item: Finish at exposed edge(s) with proprietary aluminium trim

5 Insert manufacturer and catalogue no.

Manufacturer ....

Cat. no. ....

6.5.3 Asphalt finish to external steps/landing

General

Location: ....

This clause may be used to make existing steps and landings waterproof. Granolithic screed on a suitable waterproof compound is an alternative method.

1 Item: Thoroughly clean area to be asphalted and hack surface to provide key.

2 Brickwork: Grind out 25 × 25 mm splayed chase to all upstands. Top of chase to be 150 mm above landings and minimum 75 mm above line of nosings.

2 Select either 2 or option 2.

Option 2 Brickwork: Hack off render to all upstands up to a cut line 150 mm above landings and 75 mm above line of nosings. Dub out exposed brickwork in 1:1:6 render with scribed surface to receive asphalt. Form 25 × 25 mm splayed chase with top of chase at line cut in render.

3 Asphalt: Lay 25 mm paving asphalt to BS1162 2 equal coats breaking joints top coat 15 mm additional grit. Form 150 mm upstands to landings and to 75 mm above line of nosings with 2 coat angle fillets. Tuck into chases. Chamfer nosings. Rub in clean sharp sand while still warm. Point top of chase on completion.

6.6 Built-up roofing (including associated leadwork)

6.6.1 Built-up roof finish

General

Location: ....

*The general note to **6.5.**1 applies.*

1 Item: Thoroughly clean area to be covered.

2 Brickwork: Rake out brickwork joint immediately above (but not less than 12 mm above) highest point of top of timber kerb to receive lead cover flashing. Depth 25 mm.

2 Select either 2 or option 2.

Option 2 Render: 25 mm above highest point of top of timber kerb cut, chase using a disc cutter to receive lead cover flashing. Depth 25 mm.

Option 2 Select Option 2 if the upstands have sound render.

3 Built Up Roofing: Lay .... no. layers of .... type roofing, specification code .... manufactured by .... all in accordance with their published specification and recommendations. Include for all upstands, verges and edge details. Do not bond built up roofing upstands to brickwork.

3 Insert the number of layers (2 or 3) and the type, specification code and manufacturer of the built up roofing. A 3 layer roof is better than a 2 layer one. A specification including a high performance felt is preferable. If the roof is small check with the manufacturer whether the material costs would be disproportionately expensive if a 3 layer specification were selected because of minimum order quantities.

4 Lead: Dress Code 4 lead cover flashing over all kerbs. Girth to be 150 mm. Tuck into chase 25 mm. Wedge, point and clip with 50 mm wide Code 4 lead tacks at minimum 600 mm centres. Maximum length any piece 2.1 m with ends lapped 100 mm. Vertical stepped flashing at exposed ends.

4 Other less expensive flashings can be used in place of lead.

6.7 Zinc and lead sheet roofing

6.7.1 Zinc roof finish

General

Location: ....

If the zinc finish is applied direct to flat roof decking with the insulation laid on the ceiling between the joists (as is very common), this constitutes a 'cold roof'. The cavity between deck and ceiling must be well ventilated to avoid risk of condensation. The risk may be further reduced by the use of a paint applied vapour barrier to the ceiling.
The Zinc Development Association produce leaflets concerning the use of zinc. These should be consulted.
It is inadvisable to allow run-off from an asbestos cement roof fixed with copper nails and rivets to drain on to a zinc or zinc alloy roof because of the corrosive effect of the copper on the zinc. However the Zinc Development Association say that due to the great amount of dilution, no electrolytic action is likely or is known in their experience.
It is advisable to draw a batten and drip layout. Although cross fall joints are technically possible, they are stressed by thermal movement and liable to capillary leaks.

1 Felt: Lay felt underlay to BS 747 Type 4A (ii) brown sheathing felt (No. 1 inodorous) weight 25 kg per 25 m roll, laid loose with 50 mm laps, including over kerbs (if any)

2 Zinc: Lay 0.8 mm zinc roof complete with all rolls, battens, welts, upstands and cover flashings. Note: no cross fall joints permitted except by stepped drips.

2 Select either 2 or options 2A or 2B.

Option 2A Zinc: Lay 0.8 mm zinc/titanium alloy roof, complete with all rolls, battens, welts, upstands and cover flashings. Note: no cross fall joints permitted except by stepped drips.

Option 2A Select Option 2A if an alloy roof is required rather than one of pure zinc. It is considered more durable in a polluted atmosphere than pure zinc, is easier to work and the material is more widely available.

Option 2B Zinc: Lay 0.8 mm zinc/titanium alloy strip, complete with all rolls, battens, welts, upstands and cover flashings. Note: no cross fall joints permitted.

Option 2B Select Option 2B if a roll cap roof with bays up to 9 m long is required. This option is only economic if 18 m rolls can be used with small wastage.

6.7.2 Lead roof finish

General

Location: ....

The General note concerning 'cold roofs' for Clause ***6.7.****1 applies.*

1 Felt: Lay felt underlay to BS 747 Type 4A (ii) brown sheathing felt (No. 1 inodorous), weight 25 kg per standard 25 m roll, laid loose with 50 mm laps, including over kerbs (if any).

2 Lead: Lay Code 5 lead roof complete with all rolls, battens, welts, upstands and flashings.

2 It is advisable to draw a batten and drip layout.

6.8 Natural/asbestos cement slates and concrete tiles including associated zinc and leadwork

6.8.1 Natural slate roof

General

Location: ....

Other forms of hung roofs have not been included.

1 Felt: Using 19 mm galvanised clout nails fix to joists felt underlay to BS 747 Type 1F reinforced bitumen felt, weight 15 kg per 10 m². Dress into gutters.

2 Slating: Fix sound .... slates with 38 mm aluminium nails to tilting fillets and 25 × 50 mm tanalised battens fixed with galvanised steel cut or wire nails, minimum length 55 mm. Slates to be uniform in size and colour. Slates to be laid in parallel horizontal courses. Joints between slates to be not more than 3 mm. Lap to suit pitch.
Slate size .... × ....

2 Insert either new or second hand. Sound slates 'ring' when tapped with a coin. Defective ones sound 'dead'. Insert slate size.

3 Slating: New clay ridge tiles to match existing in size or colour.

3 Select either 3 or option 3.

Option 3 Slating: Re-fix clay ridge tiles previously set aside.

Option 3 Select option 3 if the existing ridge tiles are to be re-used. Specify their Setting Aside in Clause ***6.3.****2/2.*

4 Slating: New clay hip tiles to match existing in size and colour. New galvanised hip irons.

4 Select either 4 or option 4.

Option 4 Slating: Re-fix clay hip tiles previously set aside. New galvanised hip irons.

Option 4 Select option 4 if the existing hip tiles are to be re-used. Specify their setting aside in Clause **6.3.**2/3.

6.8.2 Asbestos cement slate roof

Location: ....

1 Felt: Using 19 mm galvanised clout nails fix to joists felt underlay to BS 747 Type 1F reinforced bitumen felt, weight 15 kg per 10 m². Dress into gutters.

2 Slating: Fix using copper nails and rivets, asbestos cement slates manufactured by

. . . .

size . . . . mm × . . . . mm colour . . . . to tilting fillets and 25 × 38 mm tanalised battens fixed with galvanised steel cut or wire nails, minimum length 55 mm. Slates to be laid in parallel horizontal courses. Joints between slates to be not more than 3 mm. Lap to suit pitch as manufacturer's instructions.

2 Insert manufacturer, size and colour.

Slate sizes may be used as follows:

	mm
Pitch 25° or more	*600 × 300*
	508 × 254
	400 × 200
Pitch 20° to 25°	*600 × 300*

3 Slating: Re-fix clay ridge tiles to match existing in size and colour.

3 Select either 3 or option 3.

Option 3 Slating: Re-fix clay ridge tiles previously set aside.

Option 3 Select Option 3 if the existing ridge tiles are to be re-used. Specify their setting aside in Clause ***6.3.****2/2.*

4 Slating: New clay hip tiles to match existing in size and colour.

4 Select either 4 or option 4.

Option 4 Slating: Re-fix clay hip tiles previously set aside.

Option 4 Select option 4 if the existing hip tiles are to be re-used. Specify their setting aside in Clause ***6.3.****2/3.*

6.8.3 Concrete tile roof

Location: . . . .

General

The strength of the roof must be carefully checked if it previously had a slate covering. Concrete tiles are considerably heavier and an existing roof may require strengthening.
See Appendix.

1 Felt: using 19mm galvanised clout nails fix to joists felt underlay to BS 747 Type 1F reinforced bitumen felt, weight 15 kg per 10 m². Dress into gutters.

2 Tiling: Fix concrete tiles type . . . .

manufactured by . . . .

colour . . . . all in accordance with manufacturer's instructions and recommendations to tilting fillets and 25 × 38 mm tanalised battens fixed with galvanised steel cut or wire nails minimum length 55 mm. Tiling to be laid in parallel horizontal courses.

2 Insert type, manufacturer and colour.

3 Tiling: New ridge tiles by same manufacturer as ordinary tiles and to match in colour.
Type . . . .

3 Insert type.

4 Tiling: New hip tiles by same manufacturer as ordinary tiles and to match in colour.
Type . . . . New galvanised hip irons.

4 Insert type.

6.8.4 Lead/zinc flashings and soakers, etc. for slate/tiled roofs

Location: . . . .

General

The Lead Development Association produce a handbook 'Lead Sheet In Building' which should be consulted.
The Zinc Development Association produce leaflets concerning the use of zinc. These should be consulted.
If concrete tiles are being used the manufacturer's recommended details for flashing, etc. should be

followed. Soakers are generally not necessary with concrete tiles except when they are plain. Proprietary abutment and flashing units are available for some concrete tiles.
Detail drawings should accompany this clause unless the roof is very straightforward.

1 Lead: Fix Code 5 lead cover flashings, aprons, saddle pieces and back gutters at abutments with brickwork. Maximum sheet length 2100 mm, minimum lap 75 mm horizontally and 100 mm elsewhere. No joints permitted in back gutter. Welt top edge. Turn 25 mm into brick joints and wedge at 600 mm centres. Stepped flashings to be wedged each step. Allow for hacking out brickwork joints and repointing joints on completion.

2 Lead: Fix Code 4 preformed soakers. Length to equal slate gauge plus lap 25 mm for nailing. Width 175 mm. Nails to be large head copper 25 mm long.

2 Select either 2 or option 2.

Option 2 Zinc: Fix 0.8 mm zinc/titanium alloy soakers. Length to equal slate gauge plus lap, plus 25 mm for nailing. Width 175 mm. Nails to be galvanised clout, 25 mm long.

Option 2 Select option 2 if required. This option is not recommended for an asbestos cement slate roof because of possible contact of the zincwork to copper nails. This would lead to corrosion of the zinc.

3 Lead: Fix Code 5 lead lining to pitched valley gutter(s). Dress over and 150 mm beyond tilting fillets both sides. Maximum length 1500 mm. Lap minimum 150 mm. Copper nailed at head of each piece.

3 Select either 3 or option 3. Ensure that all necessary carpentry is specified elsewhere in Clause **6.2.**1.

Option 3 Zinc: Fix 0.8 mm zinc/titanium alloy valley gutter(s). Dress over and 150 mm beyond tilting fillets both sides. Maximum length 2500 mm lap, minimum 100 mm.

Option 3 Select Option 3 if required. Ensure that all necessary carpentry is specified elsewhere in Clause **6.2.**1.

4 Lead: Fix Code 6 lead box gutter. Maximum length 2250 mm between drips. Width equal to gutter girth plus 450 mm minimum, increase as gutter steps down. Dress over tilting fillets each side of gutter sole and each side of slope at base of slating.

4 Ensure that all necessary carpentry including the formation of drips, is specified elsewhere in Clause **6.2.**1.

5 Lead/Zinc,

5 Specify flashings, etc. not covered elsewhere, for example around dormer windows. Detail drawings are also necessary.

6.9 Repairs to existing roof coverings

6.9.1 Repairs to existing flat roof covering

Location: ....

General

This clause should be specifically written for each case.

6.9.2 Repairs to existing pitched roof covering

Location: ....

General

This clause should be specifically written for each case.

6.9.3 Re-lay valley gutter

Location: ....

1 Demolition: (strip off roof coverings—see roofing clause) Take up defective valley gutter boarding.

1 Select either 1 or option 1.

Option 1 Take up all valley boards and set aside.

2 Carpentry: Raise valley board support struts to provide steps of 50 mm at maximum 2100 c/c and fall of 18 mm in 2.00 m each section of gutter. New struts in 75 × 38 mm × 25 mm tanalised sw to make up valley width. Re-lay valley boards and make up with new sawn tanalised sw boards, with chamfered nosing to steps all nailed with galvanised nails.

2 Ensure that there are sufficient steps for the length of the particular gutter.

3 Insulation: Lay below valley board struts. 80 mm paperfaced quilt to lap with loft insulation.

4 Zinc: 0.8 mm zinc titanium sheet made up to steps and falls and dressed up side slopes 300 mm. Rolled and dripped to suit valley width. Laid on BS 747 Class 4A (ii) brown sheathing felt no. 1 inodorous. Loose lapped 50 mm following falls. Form spout to rainwater hopper with soldered folds upstanding. No cross fall joints except at steps.

4 Select either 4 or option 4.

Option 4 Lead: Lay Code 5 lead to valley made up to steps and falls and dressed up side slopes 300 mm over tilting fillet and finished single welt. Rolled and clipped to suit valley width. No cross fall joints except at steps. Laid on felt underlay to BS 747 Part 2 Type 4A (ii) brown sheathing felt (no. 1 inodorous), lapped loose 50 mm following falls. Form spout to rainwater hopper with lead burned upstands.

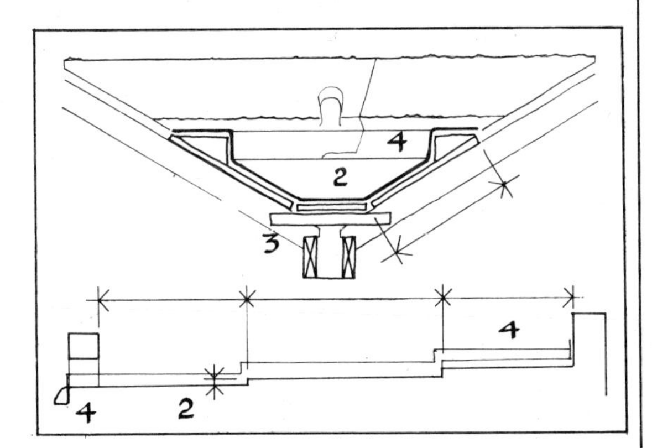

6.10 Insulation and protective finishes

6.10.1 Flat roof insulation for built-up roofing or asphalt finish (warm roof)

Location: ....

General

Although this is primarily a thermal insulation it also provides acoustic insulation.
This will provide what is known as a 'warm roof' with insulation above the vapour barrier and below the roof finish. It is important to provide sufficient thermal insulation to maintain the temperature of the vapour barrier above the dew point at all times to prevent condensation.
Unlike a 'cold roof' (where the thermal insulation is laid between the roof joists and supported by the ceiling) there is no need to ventilate the cavity between joists as long as the vapour barrier is always above dew point temperature.

1 Carpentry: 75 mm × (depth of insulation) tanalised sw batten to all abutments: 6 mm gap to abutment.

2 Built Up Roofing: Lay vapour barrier of fibre based bitumen BS 747, felt type 1B, 18 kg/10 m^2 bonded to deck in hot bitumen. Carry up upstand and allow for 450 mm overlap on top of insulation.

2 This felt is colour coded with a white stripe along the length of the roof of up to 50 mm width. Some sources recommend high performance felts.

Option 2 Insulation: Lay .... mm polyisocyanurate bitumen slabs, glass fibre faced one side, aluminium other. Hot bond to vapour barrier to glass fibre face, applying bitumen to slabs and not vapour barrier. Stagger joints.

Option 2 This is suitable for use under asphalt. A proprietary system should be specified. Insert the required thickness. 25 mm is satisfactory.

3 Insulation: Lay first layer of .... mm EHD (green stripe) polystyrene with fire retardant additive (red stripe) bonded in warm bitumen with staggered joints. Lay second layer of 19 mm fibre insulation board bonded in warm bitumen with staggered joints. Fold down bitumen felt upstand to give minimum 450 mm overlap on top of insulation.

3 This is for use under built-up roofing and is not recommended for asphalt roofs. EHD polystyrene is Extra Heavy Duty. Insert thickness of polystyrene, 25 mm is satisfactory.
The green and red stripes denoting that it is EHD and has fire retardant additive are thin pencilled lines on the edge of the sheet and are sometimes not easily found.
Many proprietary insulation slabs are available and may be specified by name, some with fibreboard factory bonded, in which case item 3 should be omitted.
Polyisocyanurate may 'gas' when heated by solar radiation and under extreme conditions. This, together with its high thermal movement could lead to the failure of the asphalt. An overlay of bitumen impregnated firbreboard as a buffer would prevent this failure. The insulation manufacturer should be consulted for their recommendation for this particular product.

4 Insulation: Expansion gap and fill between insulation and all abutments and battens to be in accordance with the manufacturer's instructions

6.10.2 Flat roof insulation for built-up roofing or asphalt finish (upside down roof)

Location: ....

General

With the insulation above the weatherproof layer, this layer and the deck are protected from extremes of thermal change. There is no risk of condensation as long as the thermal insulation is sufficient to maintain the underside of the weatherproof layer above dew point at all times. A 20% loss in thermal value may be expected for wet insulation. Acoustic insulation is also provided.

1 Insulation: Lay .... mm thick extruded polystyrene foam insulation board tight butted loose to roof covering. Joints staggered.

1 Insert thickness. 50 mm is considered satisfactory, 75 mm is better.
A proprietary system can be specified.

6.10.3 Flat roof insulation for lead or zinc roof finish ('cold' roof)

Location: ....

General

See 'General' note for Clause **6.7.**1.

1 Insulation: Wedge .... mm paper faced fibreglass quilt between all joists, paper face down. Staple paper to joists. Underside of insulation to be flush with underside of joists.

1 Insert thickness of quilt. Generally 80 mm but 100 mm or more is better. Ensure ventilation through roof voids.

6.10.4 Pitched roof insulation

Location: ....

1 Item: Lay .... mm paper faced fibreglass quilt insulation between joists, throughout roof space, paper face up, including carrying under box gutter sole (if any) and stapling to vertical studwork enclosing light wells (if any).

1 Insert thickness of quilt. Generally 80 mm but 100 mm or more is better.
An acceptable alternative to paper faced quilt is plain quilt with building paper laid over. The paper stops dust and dirt filling up the air cavities.
Ensure roof ventilation is maintained at eaves.

6.10.5 Concrete paving slabs on slab supports for trafficable 'upside down' roof

Location: ....

1 Paving: Lay 600 × 600 × 50 mm PC concrete paving on slab supports manufactured by ....

....

All cutting to be by mechanical saw. Lay 50 mm nominal 20–30 mm gravel around rain water outlet(s).

1 Insert manufacturer. Slab supports are specifically made for 'upside down' roofs and some allow easy tile levelling.
Check suitability of roof outlet.

6.10.6 Gravel finish for non-trafficable 'upside down' roof

General

Location: ....

The gravel prevents the insulation from blowing away.

1 Gravel: Lay 50 mm nominal 20–30 mm gravel overall.

6.10.7 Reflective stone chippings for non-trafficable built-up roofing or asphalt finish

General

Location: ....

For 'warm' roofs chippings are only suitable covering. Upstands and gutters may be painted.

1 Chippings: Lay 13 mm reflective stone chippings overall bedded in bitumen dressing compound manufactured by ....

.... Type ....

1 Insert manufacturer and type.

6.10.8 Reflective paint for non-trafficable built-up roofing or asphalt finish

Location: ....

1 Paint: Prepare and paint 2 coats reflective finish manufactured by ....

.... Type ....

colour ....

1 Insert manufacturer, type and colour. (Either white or silver).

7 Floors

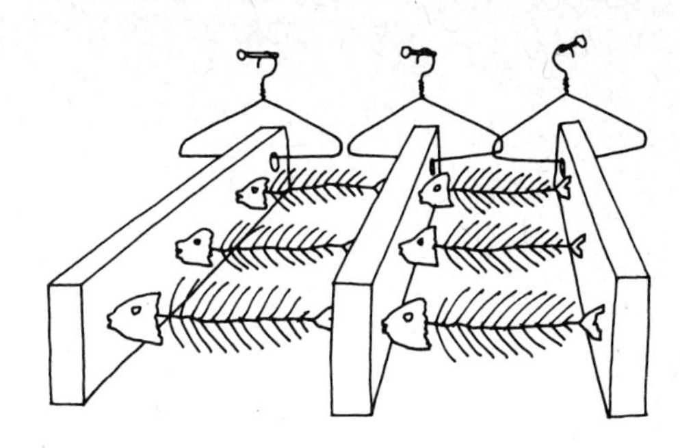

JOIST HANGERS + HERRINGBONE STRUTTING

7 Floors

7.1 Concrete ground slab

7.1.1 New concrete ground slab

Location ....

1 Excavation: To give new floor to ceiling height .... *Do not* reduce levels below top of footings without consulting the Supervising Officer.

2 Hardfill: Minimum 100 mm hardcore bed.

3 Hardfill: 25 mm blinding concrete, or 50 mm sand.

4 DPM: 1000 gauge polythene with taped joints turned up at perimeter to 150 mm above FFL. Pin behind 38 × 19 mm tanalised sw batten.

Option 4 DPM: 1000 gauge polythene with taped joints turned up at perimeter to 150 mm above FFL. Pin behind 38 × 19 mm tanalised sw batten. At abutment with timber floor, dress down between concrete and floor joists. Tape additional piece to membrane and dress up under floor covering.

5 Concrete: Mix B slab 100 mm thick cast into all fireplaces, recesses, etc. with smooth level finish.

6 Concrete: Lay 25 mm insulation board type .... manufactured by ....

....

on 1000 g DPM on Mix B slab 100 mm thick cast into all fireplaces, recesses etc. with a smooth, level finish. Insulated upstand to all perimeters to finished screed level, waterproof tape all joints, lay 65 mm thick sand/cement screed, ready to receive floor tiles.

7 Screed: 50 mm thick ready to receive floor tiles.

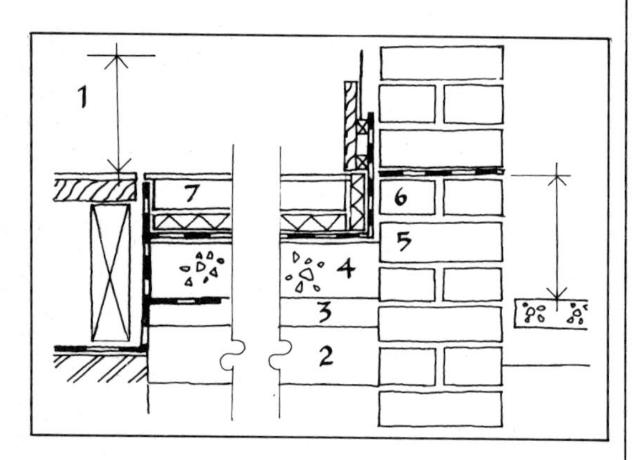

General

Local Authority standards for concrete ground slabs have been changing and the Local Authority should be consulted before specifying.

1 Insert ceiling height.

4 Select either 4 or option 4.

Option 4 Select this option where the concrete floor abuts an existing suspended floor.

6 Omit subclause 4 above. This is only one way of insulating the slab; some of the options are:
a) Insulation to edge of slab only.
b) Insulation under slab and to slab edge: slow warm-up.
c) Insulation under floor boarding (ply or chipboard): fast response but floating floor.

Use subclause 6 where slab is to be insulated and insert the type of insulation and the name of the manufacturer.

7.2 Repairs to concrete floor

7.2.1 Blinding to subfloor area

Location ....

General

Use this clause to cover earth sub-floor below existing timber floors. Insert the depth of the under

floor space to be provided. 100 mm minimum in the London Building Act and 125 mm in the Building Regulations.
This clause does not comply with regulations for new work but may be necessary to protect existing sub floor spaces.

1 Demolition: Clear out and provide a level surface to sub-floor area. Provide .... mm from finished slab level to underside of joists.

2 DPM: 1000 g polythene membrane, taped at joints turned up at perimeter 150 mm. Pin behind 30 × 19 mm tanalised sw batten.

3 Concrete: Lay 50 mm Mix A slab over DPM.

7.3 Ducts, etc. in concrete slab

7.3.1 Duct for gas service pipe

Location ....

General

Check that this clause and the length of the run conforms with current Gas Board regulations.

1 Item: Lay within hardcoro bed 100 mm vitrified clay pipes (Invert level not less than 250 mm below FFL) taken through opening cut in external wall. Pipework to be laid in a straight line between point of entry and riser position.

2 Concrete: Extra over casting concrete slab (See Concrete Slab Clause) for forming concrete access pit minimum 450 mm square at riser position. Backfill box with sand up to slab DPM level after installation of riser. Tape section of DPM to existing in slab, tape around riser and backfill to FFL in screed.

7.3.2 Duct for water main/electrical intake

Location ....

1 Item: Within hardcore bed lay 100 mm vitrified clay pipe taken through opening cut in external wall. Elbow bend flush with finished floor level. After installation of main, fill to slab level with sand and screed over.

7.3.3 Duct in concrete slab (subfloor ventilation)

Location ....

General

This clause used in conjunction with concrete slabs when there are pockets or corners under a timber ground floor where air cannot otherwise circulate. Normally at least 2 vents are required.

1 Item: Within hardcore bed lay 100 mm vitrified clay pipes as shown on drawing, with protective sand fill.

2 Brickwork: Form opening(s) in external wall and fit 150 × 225 mm terracotta airbrick(s) opposite end(s) of pipe grouted thereto.

7.3.4 Duct in concrete slab (for room ventilation)

Location ....

General

This clause provides low level natural ventilation to internal lobbies, bathrooms and WCs.

1 Item: Within hardcore bed lay 100 mm vitrified clay pipe with elbow bend flush with finished floor level.

2 Brickwork: Cut opening in external wall and fit 230 × 150 mm terracotta airbrick. Make good brickwork and grout to pipe.

2 Select either 2 or option 2.

Option 2 Item: Cut opening in external wall and provide additional vitrified clay elbow connected to horizontal pipe and to 100 mm CI stand pipe terminated with galvanised fresh air inlet hood *without* mica flap, fixed to external wall. Make good paving disturbed.

Option 2 is used when slab level does not permit the use of airbrick direct through the hall.

3 Item: Within house provide 100 mm diameter PVC pipe connected to elbow bend with further elbow to bring end flush with outside face of partition at height of .... mm above floor level. Fit 225 × 150 mm plaster louvre over end of pipe. Make good and plaster.

3 Insert height.

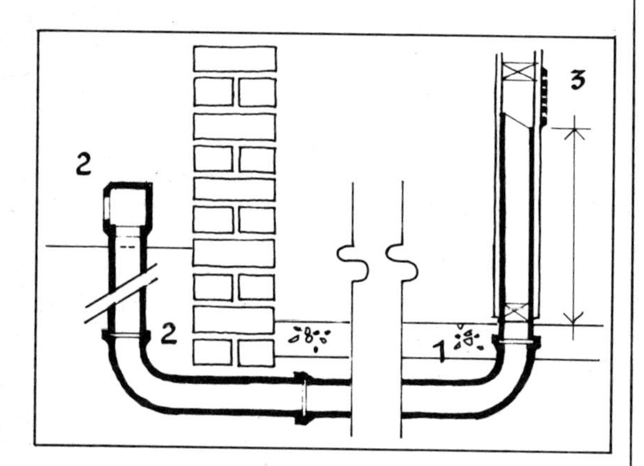

7.4 Miscellaneous ducts/chases in screed

7.4.1 Matwell in concrete slab

Location ....

1 Metalwork: 800 × 500 mm matwell frame made from 50 × 3 mm galvanised ms flat positioned in recess in screed.

2 Item: Plastic backed coconut mat to suit recess size.

7.4.2 Metal duct in screed

Location ....

1 Metalwork: Plug and screw to slab metal trunking with removable lid,

Manufacturer ....

Cat. no. .... Top of duct to finish level with screed for floor tiles to continue through. See drawing for location and length of runs.

General

This clause is to avoid making pipes inaccessible within the screed.

1 Specify the manufacturer and type of trunking to be used.
Floor tiling must also be laid out to suit removable duct covers.

7.5 New timber ground/basement floors

7.5.1 New suspended timber ground floor

Location ....

1 Excavation: To give level base for hardcore. DO NOT reduce levels below top of footings without consulting Supervising Officer.
Allow approx .... depth.

1 Minimum space between slab and underside of joists is 100 mm in the London Building Act and 125 mm in the Building Regulations. Insert depth.

2 Hardfill: Minimum 100 mm hardcore bed. 25 mm sand blinding.

2 Ensure that DPC is inserted below the level of the floor joists.

3 Concrete: Mix B slab 100 mm thick.

3 Doubling up joists or bracing floor to carry partitions etc. above is covered elsewhere.

4 Brickwork: Half brick fletton honeycomb sleeper walls at maximum 1800 mm centres. Leadcore DPC over. Height to give floor to ceiling height .... Direction of joist span as shown on drawings.

4 Subfloor ventilation is described elsewhere. Insert floor to ceiling height.

5 Carpentry: 100 × 75 mm tanalised sw wall plates bedded on sleeper walls .... × .... mm tanalised sw joists at 400 mm centres spiked to wall plates.

5 Insert joist size. Doubled up joists not covered here. Add a new subclause if required.

6 Carpentry: Ex 25 × 150 mm plain edged boarding overall.

6 Other forms of flooring can be used.

7.6 Repairs/alterations to timber ground/basement floors

7.6.1 New sleeper walls

General

Location ....

Check that tho wall and footing are suitable for the proposed loading.

1 Excavation: For foundation trench 300 mm deep.

2 Concrete: Mix A footings 300 mm wide × 300 mm deep.

3 Brickwork: Half brick fletton honeycomb sleeper wall. Leadcore DPC over. Height to equal existing.

4 Carpentry: 100 × 75 mm tanalised sw wall plate bedded on DPC and wedged up to joists.

5 Allow: .... linear metres .... mm high.

5 Insert the length and height of walls.

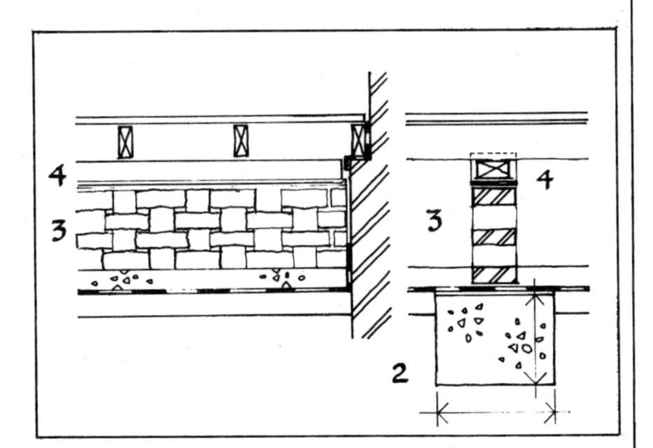

7.6.2 New sleeper piers

General

Location ....

Check that the pier and footing are suitable for the proposed loading.

1 Excavation: For foundation trench, 300 mm deep.

2 Concrete: Mix A footing 300 × 300 × 300 mm.

3 Brickwork: One brick square fletton sleeper pier. Leadcore DPC over. Height to equal existing. Wedge up to joists with slate slips.

4 Allow: .... no. piers .... mm high.

4 Insert the number and height of piers.

7.6.3 Board basement/ground floor hearth opening

Location ....

1 Demolition: Break out hearth. Clear away subfloor brickwork to general subfloor level.

1 This clause assumes that the joists run parallel with the chimney breast which is more often the case.

2 Brickwork: 2 no. one brick square fletton piers with leadcore DPC over to support trimmer/joist junction built off 300 × 300 × 300 mm Mix B concrete footing.

3 Carpentry: 500 mm × (depth of existing joists) tanalised sw noggings hung on joist hangers to hearth trimmer and chimney breast at same centres as existing joists. Lay boards to match existing adjacent staggered with adjacent.

3 Select either 3 or option 3.

Option 3 Carpentry: 50 mm × (depth of existing joists) tanalised sw noggings hung on joist hangers to hearth trimmer and back of chimney breast recess at same centres as existing joists. Lay boards to match existing adjacent staggered with adjacent.

Option 3 Use this option if hearth recess is retained.

7.7 Subfloor ventilation

7.7.1 Subfloor vents (upstand)

Location ....

General

This clause is used where the external earth level is above the level of underside of floor joists.

1 Brickwork: Cut opening in external wall, excavate externally and fit 100 mm CI standpipe and elbow built into opening and fixed to wall. Terminate with galvanised fresh air inlet hood *without* mica flap. Make good paving/ground disturbed.

2 Allow: .... no. equal spaced or where shown on drawings.

2 Insert the number of vents required.

7.7.2 Subfloor vents (airbricks)

Location ....

1 Brickwork: Cut opening in external wall and fit 225 × 150 mm terracotta airbrick.

2 Allow: .... no. equal spaced or where shown on drawings.

2 Insert the number of vents required.

7.7.3 Subfloor vents (internal walls)

Location ....

General

This clause provides for continuous cross ventilation through internal walls in sub floors.

1 Brickwork: Cut 150 × 225 mm opening through brick internal wall in subfloor space to provide through ventilation.

2 Allow: .... no. equal spaced.

2 Insert number of openings required.

7.8 New timber upper floors

7.8.1 New suspended floor

Location ....

1 Carpentry: 50 mm × . . . . mm tanalised sw joists at 400 mm centres hung on galvanised ms joist hangers to brick walls.

1 Insert the depth of tanalised sw joists required.

2 Carpentry: 38 × 38 mm tanalised herring bone strutting at . . . .

2 Insert mid-span or centre third points. It is considered good practice to insert strutting at mid-span for spans of 2.5–5.0 m and at mid-third points for spans exceeding 5.0 m.

3 Carpentry: Ex 150 × 25 mm plain edged boarding overall.

3 In areas where building regulations apply 5 mm hardboard is required on top of plain edged boarding as a smoke stop between tenancy separation floors include in floor finish schedule

4 Plasterboard: 9.5 mm Gyproc lath to soffit including 50 × 50 mm tanalised noggings at edges.

4 Select either 4 or option 4.

Option 4 Plasterboard: 2 layers 9.5 mm Gyproc lath to soffit including 50 × 50 mm tanalised noggings at edges.

Option 4 Use this option where half-hour fire resistance is required between floors.

5 Plaster: Mix C overall.

6 Cross refer: Refer to External Walls section for CP III straps.

7.9 Repairs/alterations to timber upper floors

7.9.1 Additional joists under partitions

General

Location . . . .

This clause is normally only suitable for a single storey partition. The joist supporting the partitions should be considered independently in respect of its depth and slenderness to support the new partition. Two joists are often required.

1 Carpentry: Lift floorboards as necessary. Replace on completion.

2 Carpentry: . . . . × (depth of existing joists) tanalised sw joist under line of new partition hung on joist hanger(s) to external wall(s)

2 Allowance for plaster repairs to be made in Plaster Repair Schedule.

3 Carpentry: Insert new treated sw solid bracing (section as joist) between new joist and existing joists, to line through with existing bracing/strutting.

3 Select either 3 or option 3.

Option 3 Carpentry: Insert new treated sw solid bracing (section as joist) between new joist and existing joists, and between existing joists.

Option 3 Select where there is no existing floor bracing. Normally one run of bracing at centre of span. Up to 5.0 m, 2 rows for span 5.0 m to 7.5 m.

7.9.2 Replace existing joists

General

Location . . . .

Allowance for plaster repairs to be made in Plaster Repair Schedule.

1 Carpentry: Lift floorboards as necessary. Replace on completion.

2 Demolition: Strip out defective joist.

3 Carpentry: 50 mm × (depth of existing joists) tanalised sw joist hung on joist hanger(s) to external wall(s). Make good bracing or strutting.

4 Allow: . . . . no. joists, average . . . . metres long.

4 Insert number and size of joists.

5 Cross refer: Refer to External Walls Section for CP III straps.

7.9.3 Cut back decayed joist ends

Location ....

1 Carpentry: Lift floorboards as necessary. Replace on completion. Support floors from below as required.

2 Carpentry: Cut back decayed joist to extent directed by Supervising Officer. Provide new 50 mm × (depth of existing joists) section lapped 1200 mm with existing joist and bolted thereto with 4 no. 12 mm diameter coach bolts, 50 mm steel washers and 75 mm double sided toothed plate timber connectors. Hang to external wall(s) on joist hanger(s).

3 Carpentry: Allow for cutting back and lapping .... no. joist(s).

3 Insert number of joists to be cut back and lapped.

4 Cross refer: Refer to External Walls Section for CP III ties.

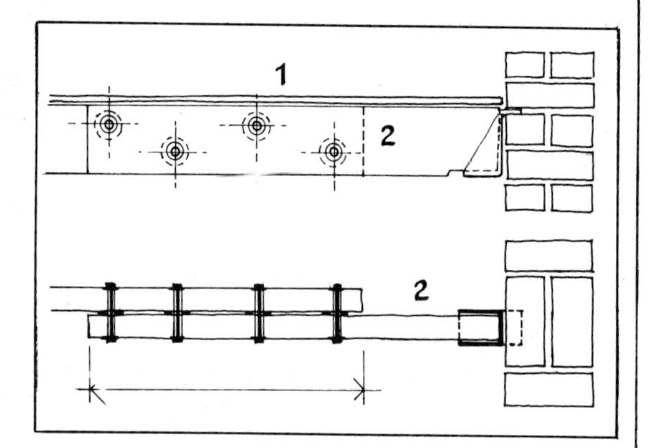

7.9.4 Board hearth opening (no recess, joists parallel to breast)

General

Plaster repair not included. Insert the area of plaster to be repaired in the Plaster Repair Schedule assuming the ceiling below is to be retained.

Location ....

1 Demolition: Break out hearth and carefully clear away brick arch below.

2 Carpentry: 1 no. 50 mm × (depth of existing joists) tanalised sw joist hung on joist hangers to existing trimmers. Lay boards to match existing adjacent, staggered with existing.

2 Existing trimmers are often suspect and it may be necessary to replace 1 no. full length together with stub joists either side of chimney breast.

7.9.5 Board hearth opening (no recess, joists right angles to breast)

Location ....

1 Demolition: Break out hearth and carefully clear away brick arch below.

2 Carpentry: 50 mm × (depth of existing joists) tanalised sw noggings hung on joist hangers to hearth trimmer and chimney breast at same centres as existing joists. Lay boards to match existing adjacent, staggered with existing.

7.9.6 Board hearth opening (recess, joists parallel to breast)

General

Plaster repair not included. Insert the area of plaster to be repaired in the Plaster Repair Schedule assuming the ceiling below is to be retained.

Location ....

1 Demolition: Break out hearth and clear away brick arch below.

2 Carpentry: At back of fireplace fix 75 × 50 mm tanalised sw bearer rawlbolted to brickwork. 1 no. 50 mm × (depth of existing joists) tanalised sw joist hung on joist hangers to existing trimmers. Lay boards to match existing adjacent, staggered with existing.

2 Existing trimmers are often suspect and it may be necessary to replace 1 no. full length together with stub joists either side of chimney breast.

7.9.7 Board hearth opening (recess, joists right angles to breast)

Location ....

General

Plaster repair not included. Insert the area of plaster to be repaired in the Plaster Repair Schedule assuming the ceiling below is to be retained.

1 Demolition: Break out hearth and clear away brick arch below.

2 Carpentry: 50 mm × (depth of existing joists) tanalised sw noggings hung on joist hangers to hearth trimmer and back of chimney breast recess at same centres as existing joists. Lay boards to match existing adjacent, staggered with existing.

7.9.8 Level existing hearth

Location ....

General

Hearths which are level and sound can normally be retained where they are likely to be covered by floor tiles or carpeting.

1 Item: Clean off existing surface to hearth and remove all loose material. Make good in 1:1:6 render and surface overall with latex screed to match adjacent floor level.

7.9.9 Level existing floor

Location ....

General

This clause only levels the floor. The ceiling below will remain out of true.

1 Demolition: Lift existing boards setting aside sound lengths for re-use.

2 Carpentry: Spike tanalised sw firring pieces to tops of existing joists to produce a level floor.

3 Carpentry: Relay existing floor boards and new floor boards as existing adjacent.
Allow .... linear metres of new boarding.

3 Select either 3 or option 3.
Insert the number of linear metres.
Adjust door in door schedule.

Option 3 Carpentry: Lay ex 25 × 150 mm plain edge boarding overall.

4 Joinery: Provide hardwood threshold(s) at door opening.

7.9.10 Sound reducing floating floor

Location ....

General

This specification may give Grade 2 sound insulation from airborne and impact sound. However the effectiveness varies considerably according to the condition of the existing structure.

1 Carpentry: Ex 62 × 19 mm sw battens with 12 mm self adhesive foam on underside laid at 400 mm centres laid .... Screw 19 mm t & g flooring grade chipboard to bearers. (*Do not* penetrate battens). Floor to float free of skirtings. Form access panels over pipe runs.

1 Insert either on top of existing joists or on top of existing floor boards over joists below. T & G flooring ply can be used.

2 Joinery: Provide hardwood threshold at door openings separate from floating floor where change of level occurs.

3 Painting: Seal with 2 coats clear poly-urethane sealer.

3 This item may be used to give the floor a self finish. Strike out and insert floor covering as required.
Adjust door in Door Schedule.

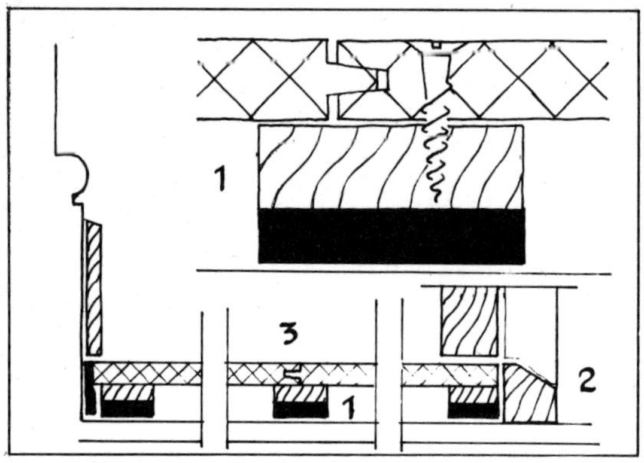

7.9.11 Floor board repairs

Location ....

1 Carpentry: Replace damaged or missing floorboards. Width and thickness to match existing adjacent.

2 Allow: .... linear metres.

2 Insert the quantity of floor boards.

3 Allow: the sum of £ .... per linear metre for the refixing of sound existing floor boards. (Contractor to insert rate).

7.10 False ceiling

7.10.1 False ceiling (tank housing)

Location ....

General

This clause is for cold water storage tanks for flats that have no access to a roof space.

1 Carpentry: 50 × 50 mm sw wall plate both sides, spiked to studs and plugged and screwed to brickwork. 100 × 50 mm sw joists at 400 mm c/c notched and spiked to bearers. False ceiling soffit to be at height .... above FFL.

1 Minimum overall height of original roof within which false ceiling for cw cistern can be accommodated is around 2800 mm (remember to allow access for removal of ball valve). Insert height required.

2 Carpentry: Trim opening for access hatch minimum 600 × 600 mm.

2 Items 2 and 4 can be omitted if some alternative form of access is required.

3 Carpentry: 12 mm shuttering grade WBP ply floor overall above joists.

4 Joinery: Ex 75 × 25 mm wrot sw lining around access opening. Ex 38 × 19 mm ogee moulded sw architrave. 19 mm blockboard loose access hatch.

5 Plasterboard: 9.5 mm Gyproc lath to soffit.

6 Plaster: Mix C overall.

7.10.2 Independent sound reducing ceiling (plasterboard)

Location ....

General

This clause is to reduce the sound level between floors or below a balcony. Joists must not be connected to floor/ceiling joists above.

1 Carpentry: .... × 50 mm tanalised sw ceiling joists at 400 mm centres below or between roof joists hung on joist hangers to brickwork. Finished floor to ceiling height to be .... mm.

1 Insert joist depth.
Insert floor to ceiling height.

2 Insulation: 100 mm sound deadening fibreglass quilt draped over joists.

3 Plasterboard: 9.5 mm × 1200 × 406 mm Gyproc lath to underside of joists including 38 × 38 mm tanalised noggings at edges.

3 Select either 3 or option 3

Option 3 Plasterboard: 2 layers 9.5 × 1200 × 406 mm Gyproc lath to underside of joists including 38 × 38 mm tanalised noggings at edges.

4 Plaster: Jute scrim to internal angles. Thin-coat metal angle bead to external angles. 5 mm Thistle board finish coat, smooth finish.

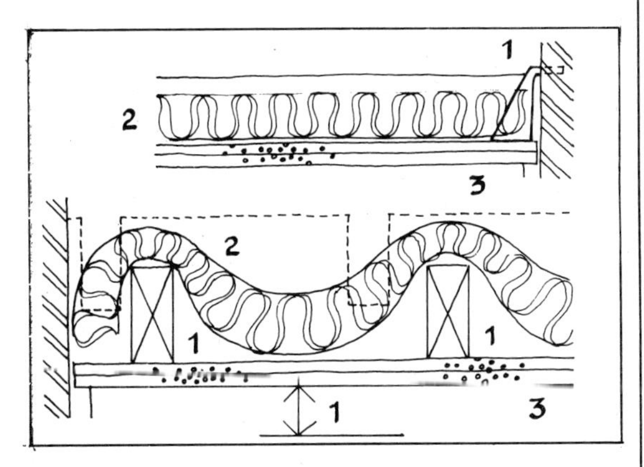

7.11 Ducts, etc. in floors/ceilings

7.11.1 Ducted air vent

Location ....

General

Some authorities prefer a separate fresh air supply direct to each bathroom or WC as well as a ventilated lobby and mechanical ventilation. Some authorities ask for a ventilated lobby with a positive airflow to bathroom via grille or similar. Check for particular Environmental Health Requirement.

1 Brickwork: Cut opening and fit 230 × 150 mm terracotta airbrick on external face of wall. Make good brickwork.

1 Select either 1 or option 1.

Option 1 Brickwork: Cut opening, take asbestos pipe to external face of wall, fit wall cowl by ....

....

to pipe, make good brickwork.

Option 1 Insert name of manufacturer.

2 Item: Between joists clip 100 mm diameter asbestos pipe grouted to airbrick or cowl. Right angle bend cut to bring end of pipe flush with ceiling.

2 Select either 2 or option 2.

Option 2 Item: Between joists clip 100 mm diameter asbestos pipe grouted to airbrick or cowl. Right angle bend and upstand where shown with further right angle bend to bring end flush with external face of partition, or bath panel.

3 Plaster: Cut ceiling and fit 230 × 150 mm plaster louvre ventilator. Make good plaster.

3. Select either 3 or option 3.

Option 3 Item: Mechanical extract fan by electrical subcontractor.

4 Carpentry: Take up floorboards as necessary, face joist each side of pipe with 9 mm fire-resistant board, cut 9 mm short of floor over. Lay 9 mm fire-resistant board over joist space sitting on fire-resistant board linings. Refix floorboards.

4 This is the appropriate clause when the vent pipe perforates a half hour fire resisting ceiling over.

5 Item: 225 × 75 mm white plastic air vent to each side of bathroom door, hole cut in door to suit.

7.12 Hatches, etc. in floors/ceilings

7.12.1 Manhole access hatch (internal manhole cover below floor level)

Location ....

1 Concrete: Shutter to form opening in 100 mm concrete slab size .... mm × .... mm. Form rebate to screed to receive 65 × 75 mm SW frame.

1 Select either 1 or option 1. Insert the overall hatch frame dimension. This option is used with a concrete slab.

Option 1 Carpentry: Trim floor joists to form opening over sub-floor manhole size .... mm × .... mm.

Option 1 This option is used where the manhole is below a suspended timber floor. Insert the size of the opening.

2 DPM: Cut back sheet DPM and lap from manhole opening up to hatch frame rebate.

3 Drainage: Refer to Drainage clauses for manhole and manhole cover specification.

4 Carpentry: 50 × 75 mm SW chamfered frame to rebate in screed, plugged and screwed to slab or screwed to floor joists. Hatch cover in 19 mm blockboard on 32 × 75 mm SW bearers at 300 mm centres screwed to blockboard.

5 Item: Aluminium trim to hatch and frame in 19 mm × 19 mm × 3.5 mm angle screwed at 100 mm ccs. flush to floor finish.

6 Ironmongery: 35 mm diameter CP flush ring rebated to hatch cover to finish flush to floor.

7 Cross refer: Refer to drawing No. ....

7 Insert drawing No.

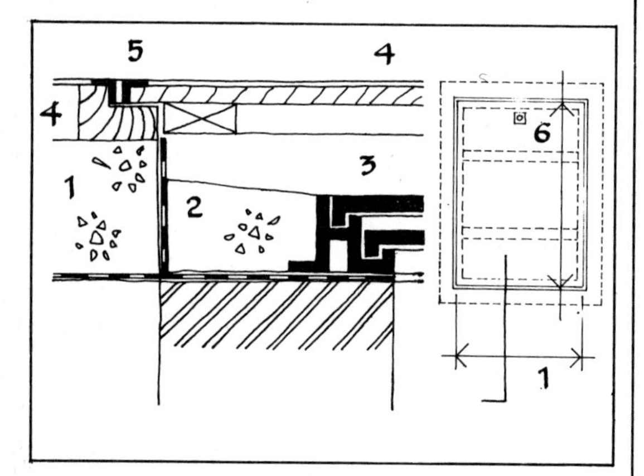

7.12.2 Floor access hatch (stop valves)

Location ....

1 Carpentry: Trim opening between adjacent floor joists to give opening approximately 300 × 400 mm. Form loose access panel of sw boarding screwed to 25 × 25 mm sw ledges resting on half width of joists.

2 Item: Trim opening and panel with 25 mm aluminium angle and fix only chrome plated flush ring handle. Angle and ring to be fitted after laying of floor finish and/or sub-flooring if any.

7.12.3 Roof void access hatch

Location . . . .

1 Carpentry: Form opening in ceiling. Trim between existing joists using 125 × 50 mm tanalised sw trimmer. Minimum dimensions of hatch 600 × 600 mm.

2 Joinery: Ex 75 × 25 mm sw lining around access opening. Ex 38 × 19 mm ogee moulded sw architrave 19 mm blockboard loose access hatch with 2 layers 9.5 mm plasterboard screwed to upper face.

3 Plaster: Make good around opening.

7.13 Bearers for water tanks

7.13.1 Bearers for central heating F/E tank

Location . . . .

1 Carpentry: Frame up stool for central heating feed and expansion tank. 75 × 50 mm tanalised sw with 12 mm blockboard base. Cross refer to Plumbing section.

7.13.2 Bearers for coldwater storage cistern

Location . . . .

1 Carpentry: 4 no. 75 × 50 mm tanalised sw bearers at 300 mm centres to span at right angles to minimum 4 no. ceiling joists. 1200 × 1200 mm 12 mm WBP ply base spiked thereto.

1 Check against size of cw cistern.

7.13.3 Firestop existing suspended floors

Location . . . .

1 Carpentry: Take up floorboards as necessary and re-lay on completion.

2 Carpentry: Trim out opening around riser pipes, line sides of opening with expanded metal strip.

3 Concrete: Mix A fill to depth of joists.

3 Select either 3 or option 3.

Option 3 Plaster: Make good plaster to ceiling below to finish level with adjacent finishes.

Option 3 May be used where fire separation is achieved by 2 layers of plasterboard.

4 Item: Provide sleeve or other joint allowing for vertical movement to all pipes.

8 Internal walls

8.1 New brick/block walls and piers (loadbearing)
8.1.1 New brick spine wall

8.2 New brick/block walls (non loadbearing)
8.2.1 New block partition
8.2.2 Blockwork duct beside chimney breast

8.3 Blockwork tanking/dry lining
8.3.1 Tanking to existing walls

8.4 Alterations/repairs to brick/blockwork walls and piers
8.4.1 Block existing opening in brickwork (lowest floor)
8.4.2 New opening in brickwork with concrete lintel (2000 mm maximum span)
8.4.3 New opening in brickwork with beam (over 2000 mm span)

8.5 New loadbearing stud partitions
8.5.1 Loadbearing stud partitions

8.6 New non-loadbearing studwork
8.6.1 Stud partition (over 2700 mm high)
8.6.2 Stud partition (under 2700 mm high)
8.6.3 Partition beside stairs
8.6.4 Stud partition to WC compartments
8.6.5 Studwork duct (full height)
8.6.6 Studwork duct (low level)
8.6.7 Pipe casing

8.7 Drylining
8.7.1 Insulating lining (stud)
8.7.2 Insulating lining (extruded polystyrene/wall board laminate)

8.8 Alterations/repairs to studwork
8.8.1 Repairing studwork with brick infill
8.8.2 Block opening in studwork with brick infill
8.8.3 Block opening in studwork
8.8.4 Form new door opening in stud with brick infill
8.8.5 Form new door opening in stud

8.9 Vertical ventilation ducts
8.9.1 Ventilation pipe

8.10 Timber trim generally
8.10.1 New skirtings
8.10.2 Skirting repairs/new skirtings to match existing
8.10.3 New dado
8.10.4 Dado to match existing
8.10.5 New picture rail

8 Internal walls

8.1 New brick/block walls and piers (loadbearing)

8.1.1 New brick spine wall

Location . . . .

General

This is a 'model' clause and should be adapted to suit each case. Structural calculations of loading will be required. The term 'spine wall' refers to an internal loadbearing wall. In terrace houses it is usually perpendicular to the party walls.

1 Demolition: Demolish existing wall. Prop structure over to satisfaction of Local Authority.

2 Excavation: Foundation trench 1000 mm deep × 600 mm wide. Backfill following casting of footing.

2 Great care must be taken not to undermine party wall footings. This should be done in close consultation with the Local Authority Inspector who may accept a shallower trench. He will require to see the bottom of the trench before the concrete is poured and may alternatively require a deeper excavation.

3 Concrete: Mix A trench fill footing 600 mm deep × 600 mm wide.

3 These are commonly used sizes but particular conditions vary. Depth and width must suit Local Authority's requirements.
Sulphate resisting cement should be used whenever there is a danger of sulphate attack to the cement.
If the new footing comes into contact with the existing party wall footings, they should be isolated from them with polythene or other movement joint material and underpinning of the party wall must be considered.

4 Cross refer: Refer to Drainage Section for support over drain runs.

4 Bridging over drains should be specified in Drainage Section.

5 Brickwork: Construct one brick fletton wall as shown on drawing no. . . . . with leadcore DPC at level of adjacent DPC and lapped with adjacent DPC by 150 mm. Bond to party walls. Pin up to existing joists/wall plate over.

5 Insert drawing no. . . . .
If a new DPC is provided for the party walls the spine wall DPC should lap with this.
For two storey houses the thickness of the wall can in some circumstances be reduced to half brick. Blockwork of suitable crushing strength may be used instead of brickwork.

6 Brickwork: Quoin up opening(s) where shown on drawings with standard PC RC lintels over.

6 Support over openings wider than 2 m will require calculated beams.

7 Plaster: Mix B render/set to both sides.

8.2 New brick/block walls (non loadbearing)

8.2.1 New block partition

Location . . . .

General

For use only at lowest level of building or on the rare upper level concrete slab.

1 Carpentry: Where partition runs between joists in direction of span, insert 100 × 50 mm tanalised noggings between floor joists at head at 600 mm centres.

2 Brickwork: 75 mm blockwork built off slab. Strength 3.5kn/m² or better. At junction(s) with existing brickwork build in galvanised wall ties in every course of blockwork. Secure at head with ms ties built into perpends and nailed to joists or

noggings in floor over at maximum 600 mm centres.

3 Brickwork: Quoin up opening(s) where shown with PC RC lintel(s) over.

4 Plaster: Mix B render/set both sides.

8.2.2 Blockwork duct beside chimney breast

General

Location ....

For use only at lowest level of building or on the rare upper level concrete slab.

1 Demolition: Cut away plaster to ceiling/wall where new duct abuts.

1 This is not necessary if new plaster is being specified to ceiling and walls and where new duct opening is being made through the floor.

2 Brickwork: 75 mm blockwork built off slab. Form opening(s) for access panel(s) where shown on drawings. PC RC lintel over.

3 Joinery: Construct access panel(s) as shown on drawing no. ....

3 Insert drawing no. ....
A fire resisting panel must be used if the duct itself is required to be fire resisting.

4 Plaster: Mix B render/set to room side.

8.3 Blockwork tanking/dry lining

8.3.1 Tanking to existing walls

General

Location ....

This clause is suitable for the internal tanking of basements or ground floor walls.

1 Demolition: Hack off plaster to wall(s) to be tanked.

2 Plaster: Dub out brickwork in Mix 1.

3 DPM: Prime wall with primer by membrane manufacturer. Offer up flexible self adhesive bitumen membrane and polythene membrane types ....
manufactured by ....
Membrane to be lapped 150 mm with slab DPM upstand and carried over slab by thickness of blockwork. Height of tanking .... mm.
Membrane to continue 150 mm beyond vertical DPC and mask bridging to wall DPC.

3 Insert membrane type and manufacturer.
Insert height.
The effectiveness of the junction between the vertical membrane and the slab DPM is critical.
In particularly damp circumstances it may be advisable to use the same membrane for the slab DPM rather than the polythene membrane specified in Clause **7.1.**1.

4 Brickwork: 75 mm blockwork built off membrane lap over slab. Secure at head with ms ties built into perpends and nailed to joists in floor over at minimum 600 mm centres. Maintain 25 mm cavity between blockwork and membrane. Fill cavity with 1:6 cement/sand mortar as blockwork raised.

4 Select either 4 or option 4.
This clause is for full height tanking.

Option 4 Brickwork: .... mm blockwork built off slab to height of .... mm above FFL. Maintain 25 mm cavity between blockwork and membrane. Fill cavity with 1:6 cement/sand mortar as blockwork raised.

Option 4 Insert thickness of blockwork and height. 50 mm can be used up to about 1 m, but 75 mm is better.
The height of the blockwork should be considered with the height of kitchen fittings, etc. A shelf at worktop height for example is of more use than one at say 300 mm. For this reason it may be desirable to raise the blockwork higher than is necessary for the membrane.

5 Brickwork: Galvanised ms cramps, plastic plug and brass screw to brickwork and build into blockwork at 900 c/c above level of membrane.

6 Joinery: Ex 19 mm sw shelf with ex 38 × 38 mm planted bullnosed nosing to master plaster.

6 Only necessary if Option 4 is used.

7 Plaster: Mix B render/set overall.

7 See Appendix for render/set mix.

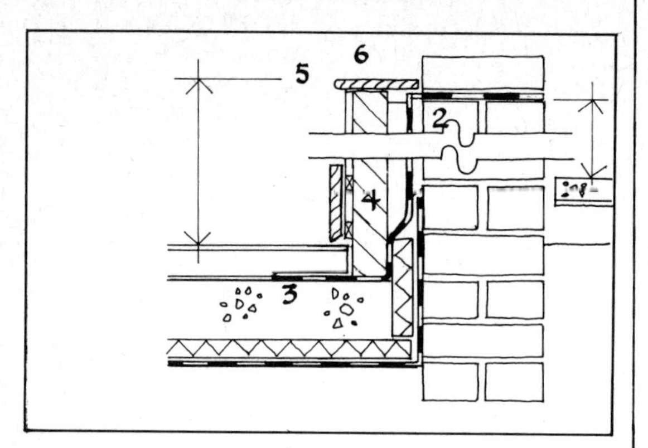

8.4 Alterations/repairs to brick/blockwork walls and piers

8.4.1 Block existing opening in brickwork (lowest floor)

Location ....

1 Demolition: Take out lintel over. Prop structure over to satisfaction of the Local Authority Inspector.

2 Excavation: For foundation trench to level of existing footings adjacent or to depth required by Local Authority.

2 This should be done in close consultation with the Local Authority. He will require to see the bottom of the trench before the concrete is poured.

3 Concrete: Mix B footing in trench 300 mm deep × 600 mm wide.

3 Select either 3 or option 3.
Depth and width must suit Local Authority requirements.
Sulphate resisting cement should be used whenever there is a danger of sulphate attack to the cement.

Option 3 Brickwork: 50 mm concrete blinding to bottom of trench. Mix A. Class B clay engineering brick spread footing to match existing adjacent, bonded to existing. Heading bond stepped 28 mm per course.

Option 3 CP 101 allows a spread brick footing for houses of not more than 4 storeys. However this option should not be used without prior consultation with the Local Authority.

4 Brickwork: Block opening .... bricks thick. Flettons above DPC, Class B clay engineering bricks below. Tooth and gauge to existing work. Leadcore DPC at level of adjacent DPC and lapped with adjacent DPC by 150 mm.

4 Insert thickness.
If a new DPC is provided to the adjacent walls the DPC should lap with this.

5 Hardfill: Backfill in hardcore.

6 Plaster: Mix B render/set to both sides.

8.4.2 New opening in brickwork with concrete lintel (2000 mm maximum span)

Location ....

General

This clause assumes the provision of a new concrete floor and so no making good to the existing floor is included.
The existing footings must be adequate to take the increased loading at the sides of the opening.
The bricks must be of sufficient crushing strength to bear the lintel.

1 Demolition: Cut opening in brickwork as shown on drawings. Prop structure over to satisfaction of Local Authority.

2 Brickwork: In .... brick wall, quoin up flush jambs in flettons. Install .... no. .... mm × .... mm standard PC RC lintels over, built in 225 mm either side of opening. 1:3 cement/sharp sand drypack over.

2 Insert thickness of wall and number and size of lintel(s). A deep opening is often spanned with several lintels.

3 Render: Mix .... render/set to quoining up to make good to adjacent plaster.

3 Insert render/set Mix A or B to match that to adjacent walls. If a large area adjacent is to be replastered 3 can be omitted and the plaster specified in the Plaster Repair Schedule.

8.4.3 New opening in brickwork with beam (over 2000 mm span)

Location ....

General

This clause must be specifically written and will require structural calculations. Guideline notes are as follows:
New openings in walls shall be carefully formed so as to ensure that no damage occurs to the building during the execution of the work. Walls and floors shall be well supported in the temporary condition and openings shall be adequately braced. Where possible the work shall be carried out in the following order, removing as little as possible of the structure at each stage:

1 Build new foundations as required.

2 Build new supports as required.

3 Insert new beams. Where two steel beams side by side are specified these should be cut away for, inserted and pinned up, one at a time.

4 Cut major opening only when new supports are in place and able to support the load.
Steel beams shall be laid on a thin mortar bed on 1:1½:3 concrete padstones unless otherwise specified. Concrete beams shall be in 1:1½:3 concrete well compacted: reinforcement shall be high yield deformed bars laid with 25 mm cover internally or 40 mm cover in faces exposed to the weather. In masonry walls beams shall be set 25 mm below the existing structure and the gap filled with 1:3 cement/sharp sand drypack, well compacted while semi-dry. In other cases beams shall be well connected to the existing structure. Existing brickwork may be used to support new beams if the stress in the wall does not exceed 0.4 N/mm² (60 psi), provided that the bricks are sound and hard, and the mortar is in good condition; in other cases the supports should be rebuilt. Timber beams, floor plates and posts shall be in GS or SS Grade material or better. Beams shall be set immediately below the floor joists which they support and be well connected to them. Beams shall be supported directly on posts which are continuous from the floor plate to existing studwork. Members shall be well dressed at joints to ensure even bearing, and shall be well nailed or otherwise connected together.

8.5 New loadbearing stud partitions

8.5.1 Loadbearing stud partitions

Location ....

General

This clause must be specifically written and will require structural calculation. Guide line notes are as follows:
When rebuilding spine walls in studwork it is generally necessary to reinstate existing diagonal bracing.
This should be connected to head and sole plates both sides, with gang nail gusset plates of which various sizes are available.
Single door openings should be formed with 100 × 100 mm sw posts and 100 × 150 mm sw lintels.

8.6 New non-loadbearing studwork

8.6.1 Stud partition (over 2700 mm high)

Location ....

1 Demolition: Cut away plaster to ceiling/wall where new partition abuts.

1 This is not necessary if new plaster is being specified to ceiling and walls.

2 Carpentry: Where studwork runs between joists in direction of span, insert 100 × 50 mm tanalised noggings between floor joists and at head at 600 mm centres.

3 Carpentry: 100 × 50 mm tanalised sw studs at 600 mm centres and behind all cut edges of laths, noggings at maximum 900 mm centres but not under lath joints. Sole, head and jamb plates. Fixing plates for all fixtures and fittings.

3 Noggings are necessary for the support of the structure and not for stiffening laths.
Studs at closer centres may sometimes be required but it is important to space them at the correct centres for plasterboard laths used.

4 Carpentry: Trim door openings with 100 × 50 mm tanalised sw heads housed to studs. Refer to door schedule for size.

5 Plasterboard: 12.7 × 1200 × 406 mm plasterboard lath both sides.

5 Select either 5 or option 5.

Option 5 Metal Lath: Expanded metal lath.

Manufacturer ....

Ref. .... both sides fixed to manufacturer's instructions.

Option 5 Insert manufacturer and reference no.

6 Plaster: Jute scrim to all internal angles, expanded metal thin coat angle bead.

Manufacturer ....

Ref. .... to all external angles. Mix C plaster to lath. Make good adjacent walls and ceiling.

6 Select either 6 or option 6A or option 6B.
For use with Clause 5.
Making good adjoining plaster is not necessary if the plaster is new.
Insert manufacturer and reference no. expanded metal lath and angle beads.

Option 6A Plaster: Jute scrim to all internal angles. Thin coat angle bead,

Manufacturer ....

Ref. .... to all external angles. Mix E plaster to metal lath. Make good adjacent walls and ceiling.

Option 6A. Insert manufacturer and reference no.
Making good adjoining plaster is not necessary if the plaster is new.
Note the dangers of having this type of plaster on site. The main contractor may be tempted to use it on external walls which are damp.

Option 6B Plaster: Mix B render/set to metal lath. Make good adjacent walls and ceiling.

Option 6B. Making good adjoining plaster is not necessary if the plaster is new.

8.6.2 Stud partition (under 2700 mm high)

Location ....

1 Demolition: Cut away plaster to ceiling/wall where new partition abuts.

1 This is not necessary if new plaster is being specified to ceiling and walls.

2 Carpentry: Where studwork runs between joists in direction of span, insert 100 × 50 mm tanalised noggings between floor joists at head at 600 mm centres.

3 Carpentry: 75 × 50 mm tanalised sw studs at 600 mm c/c and behind all cut edges of laths, noggings at maximum 900 mm c/c but not under lath joints. Sole, head and jamb plates. Fixing plates for all fixtures and fittings.

3 Noggings are necessary for the support of the structure and not for stiffening laths.
Studs at closer centres may sometimes be required, but it is important to space them at the correct centres for plasterboard laths used or considerable end noggings will be required, or the laths will have to be cut.

4 Carpentry: Trim door openings with 75 × 50 mm talanised sw heads housed to studs. Refer to door schedule for size.

5 Plasterboard: 12.7 × 1200 × 406 mm Gyproc lath both sides.

5 Select either 5 or option 5.

Option 5 Metal Lath: Expanded metal lath.

Manufacturer ....

Ref. .... both sides fixed to manufacturer's instructions.

Option 5 Insert manufacturer and reference no.

6 Plaster: Jute scrim to all internal angles. Thin coat angle bead,

Manufacturer ....

Ref. .... to all external angles. Mix C plaster to lath. Make good adjacent walls and ceiling.

6 Select either 6, option 6A or option 6B. Insert manufacturer and reference no. Making good adjoining plaster is not necessary if the plaster is new.

Option 6A Plaster: Jute scrim to all internal angles. Thin coat angle bead,

Manufacturer ....

Ref. .... to all external angles. Mix C plaster to lath. Make good adjacent walls and ceiling.

Option 6A. Insert manufacturer and reference no. Making good adjoining plaster is not necessary if the plaster is new. Note the dangers of having this type of plaster on site. The main contractor may be tempted to use it on external walls.

Option 6B. Plaster: Mix B render/set metal lath. Make good adjacent walls and ceiling.

Option 6B Making good adjoining plaster is not necessary if the plaster is new.

8.6.3 Partition beside stairs

Location ....

General

This is a 'model' clause and should be adapted to suit each case. This specification is not suitable for load bearing partitions.

1 Demolition: Take off balusters and handrail. Store handrail for refixing. Cut off nosings at edge of tread flush with string. Cut away plaster to soffit where new partition abuts.

2 Carpentry: 75 × 50 mm tanalised sw studs at 600 mm centres and behind all cut edges to laths, noggings at 900 mm centres, spaced 13 mm away from string. Sole, head and jamb plates.

2 Studs may be turned through 90° to give a 50 mm thick partition if space is restricted.
The sole may be supported on a pair of joists parallel to the string if the space under the stairs is required, say for a cupboard. The soffit of the stair must then be adequately fire proofed, generally with 2 layers of 12.7 mm Gyproc lath and Mix C.

3 Plasterboard: 12.7 mm × 1200 × 406 mm Gyproc lath both sides carried down on staircase side behind strings.

4 Plaster: Jute scrim to all internal angles. Expanded metal thin coat angle bead,

Manufacturer ....

Ref. .... to all external angles. Mix C plaster to lath. Make good adjacent walls and ceiling.

4 Insert manufacturer and reference no.
Making good adjoining plaster is not necessary if the plaster is new.

5 Joinery: Run ex 19 mm thick false string, with planted 19 × 12 mm sw staff bead, cut to match profile of stair. Ex 19 mm thick nosed sw cover plate to exposed upper face of partition.

6 Joinery: Refix handrail set aside on japanned ms brackets at centres of studs. Adapt handrail to rejoin to newel at head as agreed.

7 Cross refer: Refer to drawing no ....

7 Insert drawing no.

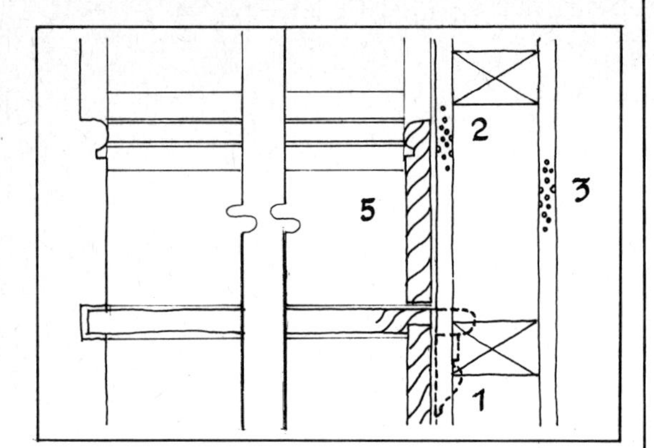

8.6.4 Stud partition to WC compartments

General

Location ....

Some authorities will accept this infill as compliance with the requirement to surround WCs with solid partitioning. It may also be used to improve sound reduction elsewhere.

1 Carpentry: Extra over Partition clause for inserting 50 mm woodwool infill in studwork around WC compartment(s) spiked to studs and finished flush with inside face of plasterboard on WC side.

8.6.5 Studwork duct (full height)

Location ....

1 Demolition: Cut away plaster to ceiling/wall where new studwork abuts.

1 This is not necessary if new plaster is being specified to ceiling and walls and where a new duct opening is being made through the floor.

2 Carpentry: 50 mm × .... mm tanalised sw framing at 600 mm centres and behind all cut edges to lath, noggings at 900 mm centres. Sole and head plates. Trim opening(s) for access panel(s) in position(s) shown on drawings.

2 Insert depth of framing. For ducts up to 2700 mm high 50 × 50 mm is suitable and for ducts above this height 50 × 75 mm.

3 Item: Firestop around services with 1:3 sand/cement mix, 50 mm thick minimum.

4 Joinery: Access panel(s) as shown on drawing no. ....

4 Insert drawing no.
The access panel(s) must have the same fire resistance as the ducts.

5 Plasterboard: 2 no. layers of 9.5 × 1200 × 406 mm Gyproc lath to face of duct.

6 Plaster: Jute scrim to all internal angles, thin coat angle bead,

6 Insert manufacturer and reference no. Making good adjoining plaster is not necessary if the plaster is new.

Manufacturer ....

Ref. .... to all external angles. Mix C plaster to lath. Make good adjacent walls and ceiling.

8.6.6 Studwork duct (low level)

Location ....

1 Demolition: Cut away plaster to walls where studwork abuts.

1 This is not necessary if new plaster is being specified to walls.

2 Carpentry: 50 × 50 mm tanalised sw framing at 600 mm centres and noggings at 900 mm centres. Sole and head plates. Trim opening(s) for access panel(s) in position(s) shown on drawings. Overall height of duct .... mm.

2 Insert height.

3 Plasterboard: 12.7 mm Gyproc lath to face of duct.

4 Joinery: Access panel(s) as shown on drawing no. .... Loose removable 12 mm blockboard shelf with sw location blocks and planted ex 38 × 38 mm rounded nosing to master top edge of duct.

Select either 4 or option 4.
4 Insert drawing no.

Option 4 Joinery: 6 mm exterior WBP ply to face of duct fixed with brass cups and screws to enable easy removal. Pencil rounded edges to all sheets.

5 Plaster: Jute scrim to all internal angles. Expamet thin coat angle bead,

5 Insert manufacturer and reference no. Making good adjoining plaster is not necessary if the plaster is new.

Manufacturer ....

Ref. .... to all external angles. Mix C plaster to laths. Make good to walls.

8.6.7 Pipe casing

General

Location ....

This clause should be specially written for each case.
Consideration should be made of pipe access, fire proofing requirements and the detailing of junctions between casing and adjoining plasterwork to avoid unsightly cracking. This may be done with sw batten fixed to the casing only to cover a plaster to timber joint.

8.7 Drylining

8.7.1 Insulating lining (stud)

Location ....

1 Demolition: Hack off existing plaster over all walls to be drylined. Cut away plaster to walls and ceiling where insulating lining abuts.

1 Hacking off may not be necessary if the plaster is a hard render and is not suffering from damp. Cutting away plaster is not necessary if new plaster is being specified to walls and ceiling.

2 Carpentry: 25 × 38 mm tanalised sw battens on tanalised packings. Fix to brickwork using galvanised ms screws into PVC plugs. Vertical battens at 600 mm centres, horizontal battens at 900 mm centres. Pack out battens as necessary to produce level finish. Horizontal battens to stop 25 mm short of verticals to allow free air movement in cavity.

3 Brickwork: Cut to size and insert plastic perpend vents to drilled out header joints in brickwork. One per bay alternately top and bottom of wall, minimum 150 mm above ground level or above injected DPC as appropriate.

3 Any airbricks specified (elsewhere) for room ventilation must be sealed from contact with the drylining cavity. Proprietary liners are available and should be specified with air bricks.

4 Plasterboard: 12.7 mm Gyproc lath, insulating grade overall.

4 The insulating lath is foil backed. The foil is not vapour proof and can therefore not act as a vapour barrier. This function is fulfilled by the plasterboard sealer.

5 Plaster: Jute scrim to all internal angles, expanded metal thin coat angle bead,

5 Insert manufacturer and reference no. Making good adjoining plaster is not necessary if it is new.

Manufacturer ....

Ref. .... to all external angles. Mix C plaster to laths. Make good to walls and ceiling.

6 Painting: 2 coats plasterboard sealer overall before final decoration

manfuactured by ....

type ....

6 Insert manufacturer and type.

7 Cross refer: Refer to drawing no. ....

7 Insert drawing no. A drawing is essential to describe this clause fully.

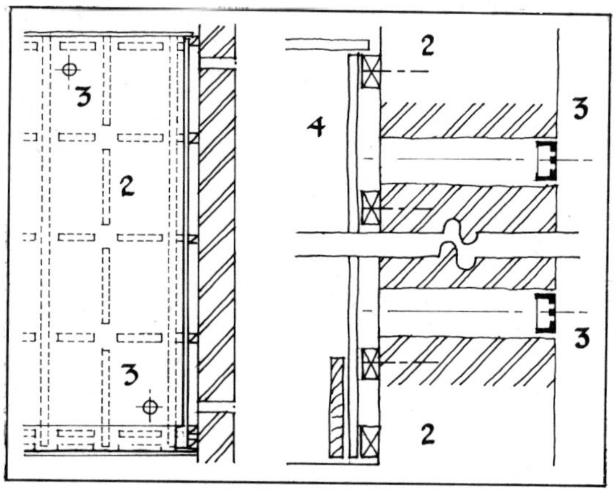

8.7.2 Insulating lining (extruded polystyrene/wall board laminate)

Location ....

1 Demolition: Hack off existing plaster over all walls to be drylined.

1 Hacking off may not be necessary if the plaster is a hard render and is not suffering from damp.

2 Render: Dub out in Mix B to all walls to be drylined.

2 Not necessary if Clause 1 is omitted.

3 Insulation: .... mm thick extruded polystyrene/wall board laminate

Manufactured by ....

Type .... fixed direct to render, all in accordance with manufacturer's instructions.

3 Insert manufacturer and type. The fixing is generally by means of a gap filling adhesive recommended by the manufacturer. Mechanical fixing is also recommeded either as a fail/safe alternative or as a secondary fixing which may be required for it to meet prevailing Fire Regulations. These must be checked before the fixing method is chosen. This clause may be amplified so that it is explicit about this point and about making good around light switches, etc. Grey finish wall board must be specified if the lining is to be plastered and cream if it is to receive direct decoration.

4 Plaster: Mix C plaster overall.

4 Select either 4 or option 4

Option 4 Plaster: Apply joint filler, joint tape and joint finish to all joints all as recommended by manufacturer.

Option 4 Use Option 4 with cream finish board for direct decoration.

5 Painting: 2 coats plasterboard sealer overall before final decoration,

Manufactured by ....

5 Insert manufacturer.

6 Cross refer: Refer to drawing no. ....

6 Insert drawing no.

8.8 Alterations/repairs to studwork

8.8.1 Repairing studwork with brick infill

Location ....

General

This model clause should be used with caution. It is often difficult to replace a segment of this type of wall without disturbing it so much that complete rebuilding becomes inevitable. Each panel of brickwork sometimes has no horizontal timber

members and bricks are often stacked on edge. If repair is attempted it is necessary to cut out plates and cut back studs both if they are rotten and when they are sound but would subsequently be below the finished level of a newly constructed solid floor. The adequacy of the existing footings should be checked.

1 Demolitions: Carefully cut out timber wall plate and studs to height of .... mm above floor level, one bay at a time. Remove loose brick infill and plaster and prop remainder over.

1 Insert height.

2 Brickwork: Build up in flettons off existing footings with leadcore DPC at level of adjacent DPC and lapped 150 mm with adjacent DPC. Pack up to carry brickwork over and adjust studs as necessary.

3 Plaster: Mix .... render/set both sides of new brickwork. Finished thickness to be same as adjacent.

3 Insert Mix A or Mix B.

8.8.2 Block opening in studwork with brick infill

Location ....

1 Carpentry: 100 × 50 mm tanalised sw soleplate at bottom of opening.

2 Brickwork: Fletton brickwork, thickness to match existing adjacent, tied to vertical studs with galvanised expanded metal strip built into brickwork every 4 courses and spiked to studwork.

3 Plaster: Mix .... render/set to both sides of new brickwork. Finished thickness to be same as adjacent.

3 Insert Mix A or Mix B.

8.8.3 Block opening in studwork

Location ....

1 Carpentry: 50 mm × (width of existing studs) tanalised sw soleplate at bottom of opening and similar vertical studs and noggings to match centres of adjacent work. New studwork wedged up to existing.

2 Plasterboard: Gyproc lath to both sides. Finished surface of plaster to line through with existing adjacent.

3 Plaster: Mix C to lath.

8.8.4 Form new door opening in stud with brick infill

Location ....

1 Demolition: Cut new opening, taking away infill up to nearest adjacent studs.

2 Carpentry: 150 × 100 mm tanalised sw lintel with a spiked birdsmouth joint to the adjacent posts. 2 no. 100 × 100 mm tanalised sw posts to support lintel on and spiked to plate. Refer to Door Schedule for door size.

2 This lintel is for a single door opening.

3 Carpentry: Carry boarded flooring through where partition removed.

4 Brickwork: Infill around opening in fletton brickwork tied to studs with expanded metal strip built into brickwork every 4 courses and spiked to studwork.

5 Metal lath: Metal lath both sides

5 Insert manufacturer and reference no.

Manufacturer ....

Ref. .... where exposed fixed to manufacturer's instructions.

6 Plaster: Mix B render/set to metal lath to both sides. Make good to adjacent plaster.

8.8.5 Form new door opening in stud

Location ....

1 Demolition: Cut new opening. IF THIS INVOLVES CUTTING A DIAGONAL BRACE DO NOT PROCEED BEFORE CONSULTING SUPERVISING OFFICER.

1 When a diagonal brace in a load bearing wall is cut by the forming of a new opening a fundamental part of the structure has been disturbed. Each case must be dealt with on its merits but the rebuilding of the partition must deal with the structural problem caused by the new opening. If in doubt consult a structural engineer.

2 Carpentry: 150 × 100 mm tanalised sw lintel birdsmouthed and spiked to the adjacent posts. 22 no. 100 × 100 mm sw posts to support lintel spiked to plate. Refer to Door Schedule for door size.

2 Check that the existing partition studs are wide enough to allow for a 100 mm deep post and lintel and reduce the specified depth if necessary.

3 Carpentry: Carry boarded flooring through where partition removed.

4 Plasterboard: Gyproc lath to both sides of studwork to sides of opening. Finished surface of plaster to line through with existing adjacent.

5 Plaster: Mix B plaster lath.

8.9 Vertical ventilation ducts

8.9.1 Ventilation pipe

Location ....

General

The proposed routes of ventilation pipes should be checked with the Local Authority.
Lined flues may be an available alternative to pipes in ducts, as may pipes in floor spaces (specified in the Flooring Section).
Proprietary fusible link fire dampers and intumescent pipe liners are available if required.
A similar specification may be used for natural ventilation of top floors but for this PVC may be used.

1 Item: 100 mm diameter asbestos pipe run vertically in duct with elbow through duct panel at high level and passing through roof with proprietary neoprene sleeved weathered slate. Terminate 600 mm above roof with 'U' bend. Securely clip pipe.

1 Select either 1 or option 1.

Option 1 Item: 100 mm diameter asbestos pipe run vertically in duct. Similar shunt branches on upper floor(s) connected to main pipe 1000 mm above shunt inlet. Elbows through duct panels at high level in each room served. Pass pipe through roof with proprietary neoprene sleeved weathered slate. Terminate 600 mm above roof with 'U' bend. Securely clip pipes.

Option 1 This option may be used to gather together ventilation pipes from separate units. Consideration of sound transfer between units should be made.

2 Item: Connect to mechanical extract fan(s) installed by electrical subcontractor.

8.10 Timber trim generally

8.10.1 New skirtings

Location ....

General

When replacement skirtings are lower than the original one a provision must be made for making good the plaster This clause is for replacement of skirtings in entire rooms. Removal of skirtings to be specified in Demolitions.

1 Joinery: Around all rooms specified run ex 25 × 150 mm sw skirtings with ex 19 mm planted staff bead including all tanalised grounds where fixed to brickwork. Prime overall before fixing. Do not pierce DPM.

1 Select either 1 or option 1A or 1B, or rewrite to be the size and shape of skirting required.

Option 1A Joinery: Around all rooms specified above, run ex 25 × 150 mm skirting with pencil rounded exposed edge including all tanalised grounds, where fixed to brickwork. Prime overall before fixing. Do not pierce DPM.

Option 1B Joinery: Around all rooms specified above, run ex 25 × 150 mm plain edge skirting including all tanalised grounds where fixed to brickwork. Prime overall before fixing. Do not pierce DPM.

2 Joinery: In the following rooms .... fix skirtings through joinery lining manufactured by .... type .... Do not pierce DPM.

2 Insert room numbers, manufacturer and type. Pretreated linings are available and advisable in the lowest level and against all walls that have suffered from damp or dry rot.

8.10.2 Skirting repairs/new skirting to match existing

Location ....

1 Joinery: .... linear metres of sw skirting to match existing in profile and mitred to existing at junctions including all tanalised grounds where fixed to brickwork. Prime overall before fixing.

1 Insert length required.

2 Joinery: To the following rooms .... fix skirtings through joinery lining manufactured by ....

type ....

2 Insert room number(s), manufacturer and type. See note clause above.

8.10.3 New dado

Location ....

1 Joinery: Around rooms specified, run ex 19 × 38 mm sw ogee moulded dado rail 1000 mm above floor level/line of stair nosings.

1 Alter the moulding type as required.

8.10.4 Dado to match existing

Location ....

1 Joinery: .... linear metres sw moulded dado to match existing in profile and mitred to existing at junctions.

1 Insert length required.

8.10.5 New picture rail

Location ....

1 Joinery: Around rooms specified, run ex 50 mm × 19 mm ogee moulded picture rail at height of .... mm above floor level.

1 Insert height. Alter the moulding type as required.

9 Internal stairs

9 Internal stairs

9.1 Internal stairs

9.1.1 New internal staircase

Location ....

1 Joinery: Construct new sw staircase with .... no. equal risers, as detail drawing no. .... Treads and risers rebated and grooved, glued, blocked and bracketed together and housed to strings. Bearers housed to newel and strings. Winders and landings cross tongued and glued. Outer string housed to newels, wall string screwed and plugged to walls. Site measure all dimensions not shown. Site treat completed staircase with 2 coats clear timber preservative before fixing.

1 Insert the number of risers and drawing number. A drawing must accompany this clause to give further detail.
Typical sizes of members are:

Newels: ex 100 mm square
Strings: ex 280 × 32 mm
Treads: ex 32 mm with round nosings
Risers: ex 25 mm
Trimmers under winders: ex 100 × 75 mm
Balusters: ex 32 × 32 mm

9.1.2 Repair existing staircase

Location ....

1 Joinery:

(a) Replace .... no. worn nosings
(b) Replace .... no. worn treads.
(c) Fix .... no. loose treads by securing wedges and refixing glued blocks.
(d) Fix .... no. loose newel posts.
(e) Fix .... no. loose carriage beams.
(f) Fix .... no. loose balusters.
(g) Replace to match existing .... no. missing balusters.
(h) Replace in ex 25 × 25 mm sw .... no. missing balusters.
(i) Restrain handrail in .... no. locations with 9 mm diameter ms ties with flattened drilled and countersunk ends screwed to handrail and staircase string.

1 Insert the number of repairs in the relevant subclauses.
This is a guideline clause only since the possible variety of repairs is wide. Do not include those which do not apply and insert additional materials as required.

9.1.3 Tubular ms newel

Location ....

1 Metalwork: 25 mm diameter 16 swg ms tube welded at base to 75 × 75 mm 16 swg ms plate holed for 4 no. c/s screws and welded at top to 25 × 75 mm 16 swg ms plate holed for 2 no. c/s screws and formed to follow line of underside of handrail. Screw in position specified.

Allow .... no.

1 Useful for strengthening existing continuous flight handrails. Insert the number of stays required.

2 Metalwork: Measure on site and provide ms stay approximately 150 mm long 6 mm round bar with flat flanges for screwing to underside of handrail and face of stair string.

Allow .... no.

2 Useful for strengthening loose handrails by bracing them to the passing string.
Insert the number of braces.
Insert number of stays to be allowed.

10 Fixtures and fittings

10.1 Kitchen fittings
10.1.1 Fix only kitchen fittings

10.2 Sanitary fittings
10.2.1 Fix only sanitary fittings

10.3 Shelving
10.3.1 Slatted shelving
10.3.2 Solid shelving

10.4 Meter cupboard
10.4.1 External gas meter cupboard
10.4.2 Internal wall mounted half hour fire resisting gas meter cupboard
10.4.3 Internal wall mounted non fire resisting gas meter cupboard
10.4.4 Internal floor mounted non fire resisting gas meter cupboard
10.4.5 External electricity meter cupboard
10.4.6 Internal wall mounted half hour fire resisting electricity meter cupboard
10.4.7 Internal wall mounted non fire resisting electricity meter cupboard

10.5 Miscellaneous items
10.5.1 Removeable panelling for plumbing unit
10.5.2 Signs

10 Fixtures and fittings

10.1 Kitchen fittings

10.1.1 Fix only kitchen fittings

General

Location ....

This clause is for simple basic kitchens using standard kitchen units.

1 Demolitions: Where floor units abut existing walls strip off skirtings and make good plaster.

2 Joinery: Plug and screw standard sink and kitchen fittings in positions shown on drawings with bottom of wall units 500 mm above worktops. Cut and fix worktops as shown.

3 Joinery: Modify standard units as shown on drawings.

3 Include if appropriate and show and describe modification clearly on drawings.

4 Joinery: Level up floor units on sw packings and provide 3 mm hardboard false skirtings along base of units, painted black.

4 Include this option if necessary when an existing floor is not absolutely level.

5 Joinery: At junctions of units and walls as shown on drawings, plug and screw ex 25 mm × (width of gap) sw spacing battens to walls to same profile of units to ensure tight junction between end of unit and wall/skirting.

5 Fitting the unit tight against the wall causes problems with the door opening against high skirtings. Spacing the end unit off the wall by the skirting thickness solves this but causes a gap between the unit and the wall. It is also necessary to specify oversized worktops to be cut to suit on site. This should be described on the drawings and would be covered by Clause 3 above.

6 Glazed Tiling: 2 courses 150 × 150 mm white glazed tiles on walls above the joint of the work tops and sink with the walls, including across cooker position. Tile behind cooker position down to skirting level. Cut tiles behind sink unit to line with adjacent worktop. Seal joints between tiles and worktop, tiles and sink and between sections of worktop with movement joint sealant

6 Insert manufacturer and type.
Using two courses of tiles allows 200 mm from the top of the tiles to the underside of wall cupboards; sufficient for cooker and/or socket outlets without the need to cut the tiles.
This is a minimum specification. The clause can, of course, be altered to include more and different tiles if desired.

Manufactured by ....

Type ....
Internal angles of tiling on studwork and junctions of studwork to solid construction to be filled with same movement joint sealant.

7 Painting: DO NOT PRICE HERE: Ensure walls inside units are painted 2 coats white emulsion to a good finish free from defects.

7 This is not necessary if the units are backed. The price should be within the Internal Decoration Schedule.

8 Cross Refer: Refer to Kitchen Layout drawing no. ....

8 Insert drawing no.

10.2 Sanitary fittings

10.2.1 Fix only sanitary fittings

General

Location ....

This clause is for a simple basic bathroom without special details.

1 Item: Chase wall plaster to accommodate bath lip(s) and make good down to bath(s).

1 This is necessary only when the bath is being inserted against walls which are not being replastered and when space is very limited. Clause 5 should not be necessary if this is done.

2 Sanitary Appliances: Fix only all items as shown on drawings. WC pans and cisterns to be fixed with brass screws.

3 Carpentry: 38 × 38 mm tanalised sw framing for bath panel(s). Fix panel supplied by others with brass cups and screws.

*3 Select either 3 or option 3.
The bath panel itself is generally specified on the Sanitary Schedule and supplied by the sanitary fittings supplier.*

Option 3 Carpentry: 38 × 38 mm tanalised sw framing for bath panel(s) including make-up pieces at end(s) clad in same material as panel. Fix panel and make-up pieces from panelling supplied by others with brass cups and screws.

*Option 3 See Note 3 above.
Make up pieces, if necessary should be shown on the bathroom detail drawings.*

4 Glazed tiling: 2 courses 150 × 150 mm white glazed tiles all round bath(s) and behind basin(s). Seal joints between fittings and tiles with movement joint compound

Manufactured by ....

Type ....
Joints at internal angles of tiling in studwork and junctions of studwork to solid construction to be filled with same movement joint filling sealant.

4 Insert manufacturer and type. The clause can, of course, be altered to include for more and different tiles if desired.

5 Glazed tiling: Quadrant tile bead all round bath(s).

5 Quadrant tile beads are necessary when a varying gap between bath and wall surface has to be covered. However they should be used with caution as they are liable to come loose if there is movement in the floor that the bath is sitting on.

10.3 Shelving

10.3.1 Slatted shelving

Location: ....

1 Joinery: ex 19 × 50 mm pencil rounded sw slats at 70 mm centres on ex 25 × 50 mm splay fronted sw bearers each side .... no. shelves spaced 375 mm apart, projecting to inside face of door lining with lowest shelf at least 750 mm from floor level and 100 mm above hot water cylinder. Shelves to be removable.

1 Insert the number of shelves required. Select height of lowest shelf.

10.3.2 Solid shelving

Location: ....

1 Joinery: 19 mm sw lipped removable blockboard shelf on ex 25 × 50 mm splay fronted sw bearers on 3 sides. .... no. shelves .... mm wide spaced .... mm apart.

1 Insert the number, width and spacing of the shelves.

10.4 Meter cupboards

10.4.1 External gas meter cupboard

Location: ....

General

British Gas prefer gas meters to be housed in a British Gas Meter Box installed in external walls so that meters may be read without the need to enter premises and the service pipe does not enter the building. The rear wall of a dustbin store or a section of new garden wall may be used where walls to an existing building are not suitable. Gas Regions will

supply these boxes free of charge when they are to be installed in an approved position.
The British Gas publication 'Gas in Housing' should be consulted as well as the Regional Board.

1 Item: Install .... no. British Gas Meter Box(es) to be obtained (free of charge) from .... Gas Board.

1 Insert number required.
Insert name of the Regional Board.

2 Cross Refer: Refer to External Walls clause for formed openings in brickwork.

2 This should either be in 'External Walls' section or 'External Works' section if in garden wall or bin store.

10.4.2 Internal wall mounted half hour fire resisting gas meter cupboard

Location: ....

General

See General Note to Clause **10.4.**1.
When it is not possible to provide an external meter position the internal positioning is severely prescribed by British Gas.
For a service entry above ground the service pipe must rise up the outside wall of the building and pass via a sleeve into the building at a point approved by British Gas and be immediately terminated at the meter position.
For service entry below ground, the service pipe must terminate at the meter position not more than 2 m inside the building. Under a solid floor the service pipe must be ducted in a way approved by British Gas and when under a suspended floor the floor void must be vented to the outside.
For service entry in high rise buildings, the meters may be positioned at spurs from a riser in a duct vented to the outside, of sufficient fire resistance and with such fire-stopping at intervals to satisfy the Local Authority Inspector. Lateral service runs not exceeding 2 m are permitted from the riser or ring main supplied from the riser.
In a building of three storeys or less the meter may be installed in a stairway or passage which forms the only means of escape providing the meter cupboard is half hour fire resisting with a self closing door. The meter may not be installed on the sole means of escape when the building is higher.
There is no fireproofing requirement elsewhere for meter cupboards.
3871 mm² (6 in.²) ventilation must be provided to each meter cupboard. This must be to the outside for fireproof cupboards.
These requirements may vary; check with the Gas Board.

1 Joinery: 19 mm blockboard cupboard with 1 no. blockboard door. All exposed blockboard edges to be sw lipped. Frame out with ex 25 × 25 mm sw. Plug and screw to wall. Line cupboard and door internally with 6 mm fire-resistant board, screwed to blockboard. Stops to be 25 × 46 mm (finished) screwed. Overall dimensions 550 × 550 × 350 mm deep. Fit ironmongery listed in Ironmongery Schedule.

1 This is a 'model' description for a high level cupboard to house a single meter. The width should be increased proportionally to house further meters.
Allow the following on the Ironmongery Schedule: 1 pair cupboard hinges: 1 no. ball catch: 1 no. 'D' handle: 1 no. door closer.

2 Joinery: Plug and screw meter board supplied by Gas Board securely to wall within cupboard.

3 Cross refer: Refer elsewhere in the Schedule of Works for fresh air ventilation to cupboard.

3 Insert Clause number. If the cupboard is fixed to an external wall the ventilation may be by airbricks in the wall. If not, ducting is necessary.

4 Cross Refer: Refer to drawing no ....

4 Insert drawing no.

10.4.3 Internal wall mounted non fire resisting gas meter cupboard

Location: ....

General

See General Notes for Clause **11.4.**2.

1 Joinery: 19 mm blockboard cupboard with 1 no. blockboard door. All exposed blockboard edges to be sw lipped. Frame out with ex 25 × 25 mm sw. Plug and screw to wall. Overall dimensions 550 × 550 × 300 mm deep. Fit ironmongery listed on Ironmongery Schedule.

1 This is a 'model' description for a high level cupboard to house a single meter.
Allow the following on the Ironmongery Schedule: 1 pair cupboard hinges: 1 no. ball catch: 1 no. 'D' handle.

2 Joinery: Plug and screw meter board supplied by Gas Board securely to wall within cupboard.

3 Cross refer: Refer elsewhere in the Schedule of Works for air ventilation to cupboard.

3 If the cupboard is fixed to external wall the ventilation may be by airbricks in the wall. if not, ducting is necessary.

4 Cross refer: Refer to drawing no ....

4 Insert drawing no.

10.4.4 Internal floor mounted non fire resisting gas meter cupboard

Location: ....

General

See General Notes for Clause **10.4.**2.

1 Joinery: ex 25 × 38 mm sw framing plugged and screwed to brickwork. Ex 25 × 19 mm sw stops. Carry across skirting in line with and to match that to chimney breast. 2 no. ex 75 × 25 mm framed doors with 6 mm ply panels and ex 13 × 32 mm panel mouldings. Ex 25 × 63 mm moulded dado rail to full width fixed to framing and front edging of top shelf. Top of plinth to be at top of skirting height. Overall height 800 mm. Depth to suit chimney breast. Fit ironmongery listed on Ironmongery Schedule.

1 This is a 'model' description for housing a gas meter within a recess between an outside wall and chimney breast.
Note that if the chimney breast is less than 300 mm deep it will not be possible to carry the skirting straight across both cupboard and breast.
Allow the following on the Ironmongery Schedule: 2 pairs cupboard hinges: 2 no. cranked barrel bolts: 1 no. ball catch: 2 no. china knobs.

2 Joinery: Plug and screw meter board supplied by Gas Board securely to wall in position shown.

3 Cross refer: Refer elsewhere in the Schedule of Works for fresh air ventilation to cupboard.

3 Insert clause number. Air brick ventilation should be specified in the External Wall Section.

4 Cross refer: Refer to drawing no ....

4 Insert drawing no.

10.4.5 External electricity meter cupboard

Location: ....

General

Area Electricity Boards prefer meters to be housed in external cupboards that may be read without the need to enter premises. The rear wall of a dustbin store or a section of new garden wall may be used where walls to an existing building are not suitable. Gas meter cupboards generally take precedence however, if space is restricted, as it is easier to accommodate electricity meters within the building than gas. Area Boards will generally supply suitable cupboards free of charge when they are to be installed in an approved position.

1 Item: Install .... no. Electricity Meter Cupboards to be obtained from .... Electricity Board

1 Insert number required.
Insert the name of the Area Board.

2 Cross refer: Refer elsewhere in the Schedule of Works for formed openings in brickwork.

2 This should either be in 'External Walls' section or 'External Works' section if in garden wall or bin store.

10.4.6 Internal wall mounted half hour fire resisting electricity meter cupboard

Location: ....

General

See General Note to Clause **10.4.**5.

When it is not possible to provide an external meter position there are restrictions on where the service head may be positioned. It is not possible to extend an existing service cable. A new cable is necessary if a changed head position is required.
In a single dwelling the service head should be adjacent to an external door and preferably on an

outside wall. This is often not possible in an existing terraced house and a common position for the head is at high level just inside the front door. An alternative allowable position is under the stair.
In multi-occupied buildings the Area Board will generally not provide a separate service cable to each consumer. Meters can either be installed in a communal area at the service head or up to 16 accessible sub-services can be run from an intake position in a communal area to meters positioned in each dwelling (15 dwellings plus landlord's supply).
For more than 15 dwellings with a single incoming service, lateral or rising mains would be necessary.
The Distribution Board should be in a communal area.
The consumer unit should be in an easily accessible position in each dwelling.

1 Joinery: 19 mm blockboard cupboard with 2 no. blockboard doors. All exposed blockboard edges to be sw lipped. Frame out with ex 25 × 25 mm sw. Plug and screw to wall. Line cupboard and door internally with 6 mm fire-resistant board, screwed to blockboard. Stops to be 25 × 46 mm (finished) screwed. Overall dimensions .... mm L, ... mm W, ... mm H. Fit ironmongery listed on Ironmongery Schedule.

1 This is a 'model' description for a high level cupboard to house meters and/or distribution boards on a fire escape path. The dimensions will depend on the number and type of fittings to be housed.
Insert sizes.
Allow the following on the Ironmongery Schedule: 1 pair cupboard hinges: 2 no. cranked barrel bolts: 1 no. ball catch: 2 no. 'D' handles: 1 no. notice saying 'Keep Doors Locked Shut' and 1 no. door closer.

2 Joinery: Plug and screw meter board supplied by Area Electricity Board securely to wall within cupboard.

3 Cross refer: Refer to drawing no ....

3 Insert drawing no.

10.4.7 Internal wall mounted non fire resisting electricity meter cupboard

General

See General Notes for Clause **10.4.**6

Location: ....

1 Joinery: 19 mm blockboard cupboard with 2 no. blockboard doors. All exposed blockboard edges to be sw lipped. Frame out with ex 25 × 25 mm sw. Plug and screw to wall. Overall dimensions .... mm L, ... mm W, ... mm H. Fit ironmongery listed on Ironmongery Schedule.

1 This is a 'model' description for a high level cupboard to house a meter, etc. in a single family house or other position where there is no requirement that the cupboard be fire resisting.
Allow the following on the Ironmongery Schedule: 2 no. pairs cupboard hinges: 2 no. cranked barrel bolts: 1 no. ball catch: 2 no. 'D' handles.

2 Joinery: Plug and screw meter board supplied by Area Electricity Board securely to wall within cupboard.

3 Cross refer: Refer to drawing no ....

3 Insert drawing no.

10.5 Miscellaneous items

10.5.1 Removable panelling for plumbing unit

Location: ....

1 Carpentry: 50 × 50 mm tanalised sw frame all round plugged and screwed to walls, ceiling and floor.

1 and 2 For use where a packaged plumbing unit is installed at the back of a cupboard used for other purposes.

2 Joinery: 3 no. 6 mm ply panels with bevelled edges fixed to frame with cups and screws. 2 no. panels to have cut out access panels, minimum 400 × 400 mm, fixed with 4 no. ms turn buttons to give access to ball valve and all stop valves.

10.5.2 Signs

Location: ....

1 Item: Supply and fix to fanlight white self adhesive numerals as house numbers,

manufactured by ....

size .... mm. Exact position to be agreed on site.

1 Insert manufacturer and size, 75, 100 or 150 mm depending on suitability.

2 Item: Supply and fix to internal flat entrance doors black self adhesive letters

manufactured by ....

size .... mm. Exact position to be agreed on site.

2 Insert manufacturer and size, 25 or 50 mm depending on suitability.

11 Plumbing

"CHASE INTO BATHROOM AND MAKE GOOD........"

11 Plumbing

11.1 General

11.1.1 General conditions

General

These plumbing clauses should be supplemented by further standards of workmanship and materials clauses together with a 'performance' specification for central heating works which is listed, where applicable, in conjunction with these clauses.

1 Water supply: The installation is to be carried out in accordance with all the materials and workmanship clauses of this specification. The contractor's attention is drawn particularly to the following points:

2 All pipework to be in copper, excepting overflows.

3 The installation to conform with Water Authority Byelaws and CP310. Pipework layout to be balanced so that lower fittings do not starve higher fittings or vicc versa.

4 Location of all pipe runs to be agreed on site with Supervising Officer *before* installation.

5 All pipework buried in concrete floors or underground to be wrapped in waterproof tape. There are to be no buried joints.

6 No notches to be cut in joists more than 1/8th of span from bearing end. Notches to be not greater than 1/6th depth of joists. Avoid notching wherever possible.

7 All pipework in roof space, under suspended ground floors or under front steps to be lagged.

8 No stop valves to be positioned behind screwed down duct covers.

9 All overflows to be run separately to point of discharge and to be self-draining.

10 Each house is to have an isolating valve accessible to all residents, in common hall or position agreed with Supervising Officer.

11 Each dwelling to have an isolating valve preferably accessible to the other residents in the house.

12 All stop valves to be accessible without need for screwdriver.

11.2 Supply pipework

11.2.1 Rising main (single family house)

Location: ....

1 Excavation: For trench average 750 mm deep between Water Board stopcock at site boundary and point of entry of rising main into building. Backfill on completion.

2 Water Supply: Lay 15 mm main connected to Water Board stopcock passed through external wall and connected to brass stop valve in accessible position agreed/shown. Pipework to BS 2871 Part I Table Y wrapped in waterproof tape before laying.

3 Water Supply: Run 15 mm rising main to cold water cistern. Tee off 15 mm branch to sink. Drain valve adjacent to entry stop valve. Stop valves adjacent to cold water cistern and sink.

4 Water Supply: Tee off 15 mm branch to central heating feed and expansion tank with stop valve adjacent.

4 Include when central heating is being installed. Not needed with some combined hot and cold water units.

11.2.2 Rising main (Multi-occupied house)

Location: ....

1 Excavation: For trench average 750 mm deep between Water Board stopcock at site boundary and point of entry of rising main into building. Backfill on completion.

1 Most Water Authorities require one stop valve accessible to all flats in a house and one stop valve at the entry point of the service to each flat unit.

2 Water Supply: Lay 22 mm main connected to Water Board stopcock passed through external wall and connected to brass stop valve in accessible position agreed/shown.
Pipework to BS 2871 Part I, Table Y, wrapped in waterproof tape before laying.

3 Water Supply: Run 22 mm rising main up to highest unit. From 22 mm main tee off 15 mm branch to each flat, with flat isolating stopcock adjacent to rising main and accessible to tenant. Continue branch to cw cistern with tee off to kitchen sink.

4 Water Supply: Additional stop valve adjacent to kitchen sink.

4 Should be included if there are long runs between rising main and sink.

5 Water Supply: Tee off 15 mm branch to central heating feed and expansion tank with stop valve adjacent (Units: ....)

5 Include when central heating is being installed. Not needed with some combined hot and cold water units.
Insert number of flats to be served.

6 Water Supply: Tee off 22 mm branch to 'Multipoint' gas fired water heater.

6 Can be reduced to 15 mm if the mains water pressure is suitable.

11.3 Cold Water Storage

11.3.1 Packaged plumbing unit

Location: ....

1 Water Supply: .... packaged plumbing units. Screw frame in position.

1 Insert the number of units.

2 Water Supply: Packaged plumbing unit

2 Insert Manufacturer and reference no.

Manufactured by ....

Ref. no. ....

11.3.2 Cold water cistern

Location: ....

1 Water Supply: 227 litre (50 gallon actual) fibreglass reinforced plastic storage cistern with lid and ball valve.

1 Can be reduced to 114 litre if not supplying hw cylinder.

2 Water Supply: Insulate cistern in roof space with 25 mm expanded polystyrene with taped joints.

2 Select either 2 or option 2.

Option 2 Water Supply: Insulate cistern in roof space with proprietary plastic covered insulation jacket.

Option 2 Insert Manufacturer and reference no.

Manufacturer ....

Ref. no. ....

11.4 Hot water storage cylinders

11.4.1 Hot water storage cistern (direct)

Location: ....

1 Water Supply: 140 litre direct hot water cylinder with 75 mm fibreglass insulation jacket in plastic cover .... no. cylinders.

1 Insert number of cylinders required.

2 Grade .... to suit head.

2 Insert grade required as below:

Grade 1 for maximum working head of 25 metres.
Grade 2 for maximum working head of 15 metres.
Grade 3 for maximum working head of 10 metres.

11.4.2 Hot water storage cistern (indirect)

Location: ....

1 Water Supply: 140 litre indirect hot water cylinder with 75 mm fibreglass insulation jacket in plastic cover .... no. cylinders.

1 Insert number of cylinders required.

2 Grade .... to suit head.

2 Insert grade required as below.

Grade 2 maximum working head of 15 metres.
Grade 3 maximum working head of 10 metres.

11.5 Down services

11.5.1 Cold water service (to fittings)

Location: ....

1 Water Supply: Run 22 mm down service with separate 15 mm branches to basin(s) and wwp(s) and 22 mm branch to bath. Stop valve adjacent to cold water cistern. Drain valve at lowest point of system in accessible position.

2 Water Supply: Separate 15 mm branch to washing machine position fixed to wall at height of 700 mm above FFL with stop valve and screwed brass plug.

3 Water Supply: Additional stop valve(s) to wwp(s).

11.5.2 Cold water feed to hot water cylinder

Location: ....

1 Water Supply: Run 22 mm feed with stop valve adjacent to cold water cistern.

11.5.3 Hot water service from packaged plumbing unit

Location: ....

1 Water Supply: Run 22 mm down service with separate 15 mm branches to basin(s) and sink and 22 mm branch to bath. Stop valve adjacent to plumbing unit. Drain valve at lowest point of system in accessible position.

2 Water Supply: Separate 15 mm branch to washing machine position fixed to wall at height of 700 mm above FFL with stop valve and screwed brass plug.

11.5.4 Hot water service from cylinder

Location: ....

1 Water Supply: Run 22 mm expansion pipe to 305 mm above cold water storage cistern and loop over top. Tee off 22 mm down service with separate 15 mm branches to basin(s) and sink and 22 mm branch to bath. Stop valve adjacent to hot water cylinder on down service branch.

2 Water Supply: Separate 15 mm branch to washing machine position fixed to wall at height of 700 mm above FFL with stop valve and screwed brass plug.

11.5.5 Hot water service from multi-point instantaneous water heater

Location: ....

1 Water Supply: Run 22 mm down service with separate 15 mm branches to basin(s) and sink and 22 mm branch to bath. Drain valve at lowest point of system in accessible position.

2 Water Supply: Separate 15 mm branch to washing machine position fixed to wall at height of 700 mm above FFL with stop valve and screwed brass plug.

2 Not all 'Multipoints' are suitable for use with washing machines.

11.6 Overflows

11.6.1 Overflow from storage cistern

Location: ....

1 Water Supply: Run 19 mm overflow(s) in PVC with solvent weld joints to fall from storage cistern or packaged plumbing unit to discharge through external wall in position shown/agreed.

11.6.2 Overflow from storage cistern (combined waste/overflow manifold)

Location: ....

1 Water Supply: Run 19 mm overflow(s) in PVC with solvent weld joints to fall from storage cistern or packaged plumbing unit connected to combined waste/overflow manifold in bath.

11.6.3 Overflow from WWP

Location: ....

1 Water Supply: Run 19 mm overflow in PVC with solvent weld joints to fall from WWP and to discharge through external wall in position shown/agreed.

11.6.4 Overflow from WWP (combined waste/overflow manifold)

Location: . . . .

1 Water Supply: Run 19 mm overflow in PVC with solvent weld joints to fall from WWP and to be connected to combined waste/overflow manifold in bath.

Summary of Schedules

Job: | **Clause:** 1

External Door Schedule | **Rev:** 2

REFER TO MATERIALS & WORKMANSHIP CLAUSES: SUM UP AND CARRY TOTAL TO SUMMARY SHEET: PRICE EACH DOOR & PRICE ALL THE RATES COLUMN. 3

Key:
m/g: make good; t.m.e: to match existing; h/w: hardwood; s/w: softwood; GWPP: Georgian wired polished plate; treat (ed): apply 2 brush coats approved clear preservative; s/o: supervising officer

	Door numbers:													**Rate**	4
Repairs to existing	Remove ironmongery & m/g														5
	Remove existing door														
	m/g existing door & re-hang														
	m/g existing frame														
	New h/w cill t.m.e. housed to frame														
	Insert water bar to h/w cill														
	Remove fanlight & m/g														
	Renew glazing bars														
	New s/w w/board: brass screw & glue														
	Repair architraves														
	Replace architraves with 25mm × 75mm s/w ogee														
	Draught excluder type														16
	Replace fanlight t.m.e.														
															18
New frames/doors	Site measure width & height														
	Refer to drawing number														23
	New external frame & h/w cill by ex 63 × 88 frame ex 63 × 150 h/w cill inc. water bar														
	As above including top hung fanlight														
	New door ref: size:														26
	New door ref: size:														
	New door t.m.e.														
Linings	Fix tan. battens: 12mm WBP ply reveals @ top & sides: treat														
	Ex 38mm cill board: treated & rebated to cill														
	Ex 25 × 75mm ogee architrave														
Glazing etc.	Hack out & reglaze 6mm GWPP glass														
	Hack out & reglaze 5mm laminated glass														
	Hack out & reglaze 4mm clear glass														
	Glaze 6mm GWPP glass														
	Glaze 5mm laminated glass														
	Glaze 4mm clear glass														
	Renew perished putty renew sprigs														
	Ventilator type														40
	Point all round in mastic														
															44
															45
	PRICE HERE & CARRY FORWARD TO SUMMARY SHEET														**Total**

1 Insert job name and number. Insert clause number to follow from last clause.

2 The schedule may be detached from the rest of the specification.

3 General Note: The Appendix contains guidance showing what works are to be included by the main contractor for each of the abbreviated headings below.

4 Insert the door numbers from the working drawings.

5 The main contractor is to insert the rate for on operation of the kind noted. This is important for the pricing of on-site variations.

16 Insert the manufacturer and catalogue number of the required draught seal.

18 Insert in spare columns any special repairs not covered in the typical repair columns above.

23 Insert the drawing number.

26 Insert the reference number and manufacturers of door and door size.

40 Insert ventilator type.

44 Insert particular repair or new work reference numbers dealt with in typical abbreviated items above.

45 The main contractor is to insert the total cost of each door and the total for all doors.

Job:	**Clause:**	1
Internal Door Schedule	**Rev:**	2

REFER TO MATERIALS & WORKMANSHIP CLAUSES: SUM UP & CARRY TOTAL TO SUMMARY SHEET: PRICE EACH DOOR & PRICE THE RATES COLUMN: 3

Key:
m/g: make good; t.m.e.: to match existing; GWPP: Georgian wired polished plate; f.r.: fire resisting; f.f.: fixed fanlight; smw: site measure width; smh: site measure height

	Door Number													**Rate**
Repairs to existing	Remove existing door & set aside for reuse elsewhere													
	Remove ironmongery & m/g													
	Repair frame/lining by piecing in													
	Strip off hardboard panelling & m/g													
	Repair door by piecing in													
	Plasterboard pack panels room side 6mm f.r. board screwed 230mm c/c with 6mm quadrant lipping													
	Remove panel beads room side: fix 6mm f.r. board over panels with panel beads													
	Rehang selected door: m/g													
	Hack out & reglaze door in 6mm GWPP													
	Hack out & reglaze fanlight in 6mm GWPP													
	Hack out & reglaze fanlight in 4mm glass													
	Strip off stops													
	Repair architraves													
	Replace architrave with 25 × 75 sw													
New doors	457 × 1981 ply faced flush door 35mm													
	533 × 1981 ply faced flush door 35mm													
	610 × 1981 ply faced flush door 35mm													
	686 × 1981 ply faced flush door 35mm													
	762 × 1981 ply faced flush door 35mm													
	838 × 1981 ply faced flush door 35mm													
	914 × 1981 ply faced flush door 35mm													
	smw × smh ply faced flush door													
	Fire doors minimum thickness 44mm													
	533 × 1981 ply faced ½hr fire door													
	610 × 1981 ply faced ½hr fire door													
	686 × 1981 ply faced ½hr fire door													
	762 × 1981 ply faced ½hr fire door													
	838 × 1981 ply faced ½hr fire door													
	914 × 1981 ply faced ½hr fire door													
	smu × smh ply faced ½hr fire door													
New frames/linings/etc.	34 × 135mm s/w linings													
	34 × 110mm s/w linings													
	34 × smw mm s/w lining													
	34 × 135mm lining with f/f over													
	34 × 110mm s/w lining with f/f over													
	34 × smw mm s/w lining with f/f over													
	15 × 46mm pinned stops													
	25 × 46mm screwed stops													
	4mm clear glass to f/f													
	6mm GWPP glass to f/f													
	ex 25 × 75mm Ogee s/w architraves													
	ex 19 × 150mm chamfered h/w threshold													
	Central mullion to fanlight													
	PRICE HERE & CARRY FORWARD TO SUMMARY SHEET													**Total**

1 Insert job name and number. Insert the clause number to follow on from last schedule.

2 Insert revision number if necessary.

3 General Note: The Appendix contains guidance showing what works are to be included by the main contractor for each of the abbreviated headings below.

4 Insert the door numbers from the working drawings.

5 The main contractor is to insert the rate for one operation of the kind noted. This is important for future pricing of on site instructions.

10 This description may require a detail drawing depending upon the nature of the doors. Normally used where it is desired to maintain the appearance of the joinery and where the door edges etc. are of the required thickness of wood to comply with the Fire Regulations.

54 The main contractor is to insert the total cost of each door and the total cost for all doors.

Job:	**Clause:**	1
Window Schedule	**Rev:**	2

REFER TO MATERIALS & WORKMANSHIP CLAUSES: SUM UP & CARRY TOTAL TO SUMMARY SHEET: PRICE EACH WINDOW & PRICE THE RATES COLUMN: 3

Key:
m/g: make good; t.m.e.: to match existing; h/w: hardwood; s/w: softwood; GWPP: Georgian wired polished plate; treat: apply 2 brush coats approved clear preservative

	Window Numbers:													**Rate**	4
Repairs to existing	Remove box frame for s/o inspection														5
	Remove ironmongery & m/g														
	Treat box frame: prime: refix														
	New h/w cill t.m.e. housed to frame														
	Repair box frame														
	Replace box frame														
	Set aside sashes														
	Replace/install rat tails														
	Replace parting beads t.m.e.														
	Replace staff beads t.m.e.														
	Ease & adjust sashes														
	Adjust weights with make-weights														
	Replace lower sash t.m.e.														
	Replace upper sash t.m.e.														
	New polypropylene sash cords														
	Screw fix sashes shut														
	Draught excluder type														22
	Repair architraves														
	Replace architraves with ex 25 × 75mm s/w ogee														
	Screw fix shutters 'open'														
	Replace back linings in 3mm WBP ply														
	Replace sash pulleys (no.)														
New windows	Site measure width & height														
	New sash window to BS644 Pt.2. with h/w cill & WBP ply back linings														
	Ditto with projecting h/w cill														
	Refer to drawing no.														33
	Refer to drawing no.														
	Refer to drawing no.														
	Refer to drawing no.														
	Manufacturers Ref:														37
	Manufacturers Ref:														
	Rooflight Ref:														
	Rooflight Ref:														
Linings	Fix tanalised battens, 12mm WBP ply reveals @ top & sides: treat														
	Ex 38mm nosed cill board: treat														
	Ex 25 × 75mm s/w ogee architrave														
Glazing etc.	Hack out & reglaze 4mm clear glass														
	Hack out & reglaze 4mm obscure glass														
	Glaze 4mm clear glass														
	Glaze 4mm obscure glass														
	Glaze 5mm laminated glass														
	Ventilator type														51
	Point all round in mastic														
															54
	PRICE HERE & CARRY FORWARD TO SUMMARY SHEET														**Total**

1 Insert the job name and number. Insert the clause number to follow on from the last schedule.
2 Insert the revision number if necessary.

3 General Note: The Appendix contains guidance notes showing what notes are to be included by the main contractor for each of the abbreviated headings below.

4 Insert the window numbers.
5 The main contractor is to insert the rate for one operation of the kind noted.

22 Insert the specification required.

33 Insert the window drawing number.

37 Insert the manufacturer's reference number.

51 Insert the ventilator specification.

54 The main contractor is to insert the cost of each window and the total cost for all windows.

Job:	**Clause:**	1
Plaster Repair Schedule	**Rev:**	2

REFER TO MATERIALS & WORKMANSHIP CLAUSES. Plaster repair allowances on this Schedule are PROVISIONAL. Consult with S/O on site before hacking off existing plaster. Measure plaster repairs with SO/QS before covering up.
PRICE ALL PLASTER REPAIR TYPE RATES 3

Type Code	**Specification of Plaster Repair Types** Rates are to include all labours	**Rate/m²**
A1	Hack off plaster to solid walls. Plaster in Mix A. (Protim Premix No. 5/Sirapite).	
B1	Hack off plaster to solid walls. Plaster in Mix B. (1:2:9 cement:lime:sand/Sirapite).	
B2	Hack off plaster & lath one side of stud walls. Fix Expamet lath complete with 50 × 50mm tan s/w noggings at edges & bearers for fittings. Plaster in Mix B.	
C1	Hack off plaster & lath one side of stud walls. Fix 50 × 50mm tan s/w noggings at 900mm c/cs and bearers for fittings. Fix 9.5mm plasterboard and set in Mix C. (Thistle Board finish).	
C2	Hack off plaster & lath to both sides of stud walls. Fix 50mm × (depth of existing) tan s/w noggings at 900mm c/cs & all bearers for fittings. Fix 9.5mm plasterboard both sides and set in Mix C.	
C3	Hack off plaster & lath to ceiling. Fix 50 × 50mm tan s/w noggings between joists at all edges of plasterboard. Fix 9.5mm plasterboard & set in Mix C.	
C4	Hack off plaster & lath to ceiling. Fix 50 × 50mm tan s/w noggings between joists at all edges of plasterboard. Fix 9.5mm foil-backed plasterboard & set in Mix C.	
C5	Hack off plaster & lath to ceilings. Fix 50 × 50mm tan s/w noggings between joists at all edges of plasterboard. Fix 2 layers 9.5mm plasterboard & set in Mix C.	
C6	Fix 9.5mm plasterboard to existing plastered ceiling and set in Mix C.	
D1	Hack off plaster only to stud walls. Repair lathing and plaster in Mix D in patch (Carlite Bonding/Carlite Finish).	
D2	Hack off plaster only to stud walls. Repair lathing and plaster in Mix D in patch.	
D3	Apply dubbing out coat to new plasterboard to existing timbers to give true surface for setting coat (Carlite Bonding).	
F1	Remove defective areas of existing setting coat. Apply bonding agent & reset in Mix E (Thistle Board Finish).	

Schedule of Quantities of Plaster Repairs by Room

Type										**Total/m²**	**Price**
A1											
B1											
B2											
C1											
C2											
C3											
C4											
C5											
C6											
D1											
D2											
D3											
F1											

19 31

PRICE HERE AND CARRY FORWARD TO SUMMARY SHEET £ 32

1 Insert job name and reference. Insert clause number to follow from last schedule.

2 Insert revision number if necessary.

3 The main contractor is to price the rates for each plaster type and associated work. Plaster mixes are described in the appendix.

19 Fill in the room numbers.

31 Insert the quantities for each plaster type room by room and total in the total m² column.

32 The main contractor is to insert the total price for each plaster type and total sum for all quantities.

Job:	Clause:	1
Internal Decorations Specification	**Rev:**	2

Carry out internal decorations throughout as per Specification.
Notes: (a) REFER TO MATERIALS & WORKMANSHIP CLAUSES (b) Colours to BS4800 will be selected & instructed during contract : maximum no. 12. (c) Paint manufacturer(s)
(d) Include for all exposed pipework & radiators. 3

Type	Specification	Rate/m²	
GL/BW	BARE WOODWORK: prepare, acrylic primer, 1 undercoat, 1 gloss finish coat		5
GL/PW	PAINTED OR PRIMED WOODWORK: prepare, patch prime, 1 undercoat, 1 gloss finish coat		
EG/P	EGGSHELL TO BARE PLASTER: prepare, alkali resistant primer, 2 eggshell coats		
EG/L	EGGSHELL ON LINING PAPER: prepare, alkali resistant primer, lining paper (600 lb/ream), 2 eggshell coats		
EG/WC	EGGSHELL ON CHIP PAPER: prepare, alkali resistant primer, woodchip paper (Grade RD), 2 eggshell coats		
EM/P	EMULSION TO BARE PLASTER: prepare, matt emulsion mist coat, 2 matt emulsion finishing coats		
EM/L	EMULSION ON LINING PAPER: prepare, lining paper (600 lb/ream), 2 emulsion finishing coats		
EM/WC	EMULSION ON CHIP PAPER: prepare, woodchip paper (Grade RD), 2 emulsion finishing coats		
WP/P	WALLPAPER TO BARE PLASTER: prepare, hang selected paper (list retail price £ . /roll)		13
AP/P	ANAGLYPTA PAPER TO BARE PLASTER: prepare, alkali resistant primer, hang selected anaglypta paper (list retail price £ . /roll). 2 gloss finishing coats		14
BO	BURNING OFF: burn off paint etc. to general surfaces		
GL/BM	GLOSS ON NEW METAL: prepare, prime to suit metal, 1 undercoat & 2 gloss finishing coats		

Schedule of Room Finishes

Room	Walls	Ceilings	Woodwork New	Woodwork Old	Burn Off	Price	
							18

PRICE HERE & CARRY FORWARD TO SUMMARY SHEET £ 30

1 Insert the job name and reference. Insert the clause number to follow on from last schedule.

2 Insert the revision number if necessary.

3 Specify the manufacturer/s or paint to be used.

5 The main contractor is to insert the rate/m².

13 Insert the cost per roll of the wallpaper to be supplied.

14 Insert the cost per roll of the Anaglypta paper to be supplied.

18 Insert the room numbers in the left hand column and the abbreviated reference for each decoration operation under the relevant element. If a blank row is left between each room no. it can be used later to insert the colour chosen for each room. The main contractor is to price each room in the right hand column.

30 The main contractor is to sum up the total cost of internal decoration in the right hand box.

Job:	Clause: 1
External Decoration Specification	**Rev:** 2

Carry out external decorations throughout as per Specification.

Notes:
(a) REFER TO MATERIALS & WORKMANSHIP CLAUSES
(b) Colours to BS4800 will be selected during contract and added to this Schedule: maximum number of colours 5 (five).
(c) Paint manufacturer(s):
(d) Fillers and stoppings: the ONLY approved materials are: 3

Fill defects and rub down

	Surface	:	Specification	Rate/m² 4
1	BARE WOODWORK	:	Prepare, acrylic primer, 2 undercoat, 1 coat gloss finish	
2	PAINTED OR PRIMED WOODWORK	:	Prepare, patch prime, 2 undercoat, 1 coat gloss finish	
3	BARE IRON OR STEEL	:	Prepare, red lead primer, 1 undercoat, 1 coat gloss finish	
4	PAINTED OR PRIMED IRON OR STEEL	:	Prepare, patch prime, 1 undercoat, 1 coat gloss finish	
5	BITUMEN PAINTED IRON OR STEEL	:	Prepare, tar-sealing primer, 1 undercoat, 1 coat gloss finish	
6	GALVANISED IRON OR STEEL	:	Prepare, calcium plumbate primer, 1 undercoat, 1 coat gloss finish	
7	BARE RENDER	:	Prepare, ( ) stabiliser or similar, 2 coats ( )	11
	OR	:	Prepare, 1 undercoat, 1 coat gloss finish	
8	PREVIOUSLY PAINTED RENDER	:	Prepare, ( ) stabiliser or similar, 2 coats ( )	12
	OR	:	Prepare, 1 undercoat, 1 coat gloss finish	
9	HARDWOOD	:	Prepare, 3 coats linseed oil	
10	BURN OFF	:		13

Location/item	Operation (1–10)	Price
		15
		26
PRICE HERE & CARRY FORWARD TO SUMMARY SHEET	**Total**	27

1 Insert the job name and reference number. Insert the clause number to follow on from previous schedule.
2 Insert the revision number if required.

3 Specify the manufacturer/s or paint to be used.
Insert the type of filling and stopping to be used.

4 The main contractor is to insert the rate/m² for each surface to be decorated and the price overall for these surfaces throughout.

11 Insert the manufacturer and catalogue number of each coating to suit circumstances.

12 Insert the manufacturer and catalogue number of each coating to suit circumstances.

13 List here the areas to be burned off.

15 Insert the operations required at each location or item.

26 The main contractor is to insert the price for each location or item.
27 The main contractor is to insert the total cost of decoration.

Job:	Clause:	1
Floor Finishes Schedule	Rev:	2

Carry out floor finishes as per Schedule below:
REFER TO: MATERIALS & WORKMANSHIP CLAUSES

Sub Floor	Specification	Rate	m/²	
Hardboard	*Smokestop*: lay 4mm hardboard soaked before use, fixed with staples at 150mm centres (NOT FOR USE UNDER TILES)			5
Ply	*Below floor tiles*: lay 3mm WBP bonded ply fixed at 300mm centres with lost head nails: well punched in			
Sand	Punch nails as necessary and sand down existing floorboards to produce a clean level surface suitable for sealing			
Finish				
Tiles	Lay 2mm thick tiles with adhesive to tile manufacturers approval			10
Sheet	Lay 2mm thick sheet flooring, with adhesive to flooring manufacturers approval			11
Carpet				12
Seal	Seal new or sanded floor with 2 coats clear polyurethene floor sealer MATT/GLOSS			13
Trim				14
Aluminium Nosing	Fix aluminium safety tread nosing with grey non-slip insert to all treads			
Rubber Nosing	Fix black rubber nosing to all treads			
Threshold	Fix aluminium threshold in doorway			

Schedule of Floor Finishes

Room	Sub Floor	Finish : Colour	Trim : No	Price		
						20
						21

PRICE HERE & CARRY FORWARD TO SUMMARY SHEET £ ______ 30

1 Insert job name and reference number. Insert the clause number to follow on from previous schedule.
2 Insert revision number if required.

5 The main contractor to insert the rate per m² for each type of sub-floor.

10 The main contractor to insert the rate per m² for each finish type, insert tile type.

11 Insert sheet type.
12 Insert carpet manufacturer/type.

13 Delete as required.
14 The main contractor is to insert the rate per m² for each trim type.

20 Insert the room numbers and specify against each the sub-floor finish and trim required.
21 The main contractor to sum up and price each room.

30 The main contractor is to insert the total price for all rooms.

Job:	Clause:	1
Ironmongery Schedule	Rev:	2

REFER TO MATERIALS & WORKMANSHIP CLAUSES: SUM UP & CARRY TOTAL TO SUMMARY SHEET: PRICE EACH ITEM: MAIN CONTRACTOR TO PRICE FOR FIX ONLY. SUPPLY AND FIX. 3

	Door & Window Numbers													Rate	4
Hinges	89mm sheradised butts														5
	89mm steel butts														
	89mm steel rising butts														
	89mm steel parliament butts														
	32mm steel butts														
	356mm galvanised medium tee														
Locks, Latches, Catches	Cylinder rim night latch														11
	Deadlocking cylinder rim night latch														
	Cylinder pull														
	5 lever mortice lock with snib														
	75mm 3 lever mortice lock (57mm back set)														
	Rebate conversion set														
	F31 deadlock (meter cupboards)														
	75mm tubular mortice latch (60mm back set)														
	Rebate conversion set														
	Mortice latch (100mm back set)														
	Rebate conversion set														
	Bathroom mortice indicator bolt														
	Nylon roller catch														
	Magnetic catch (8lb pull)														
	190mm galvanised suffolk latch														
	Galvanised automatic gate catch														
Handles Knobs	Lever latch furniture														27
	Escutcheons for above (pair)														
	White china furniture														
	Black iron centre knob														
	100mm 'D' handle														
Door Sundries	100mm door chain														32
	75mm barrel bolt														
	75mm cranked barrel bolt														
	150mm flush bolt														
	150mm cabin hook														
	Black iron letter plate														
	Internal letter flap														
	Door letters/numerals														
	Black rubber door stop														
	Coat hook														
	Notice 6mm letters "TO BE KEPT SHUT"														
	Door closer type:														44
	Door Closer type:														45
															46
Windows	Sash lifts : pair														47
	Galvanised sash pulls (100mm) : pair														
	"Brighton" sash fastener														
	Sash screw														
	Acorn sash stop														
	250mm casement stay														
	Casement fastener														
	"Easiclean" casement hinges														
	Window manufacturers set														56
	PRICE HERE & CARRY FORWARD TO SUMMARY SHEET														**Total**

1 Insert the job name and number. Insert the clause number to follow on from the last schedule.

2 Insert Revision number if necessary.

3 Insert the supplier of the required ironmongery and add any other reference to quotations for ironmongery, etc. Delete 'fix only' or 'supply and fix' as required.

4 Insert door numbers.

5 Insert the catalogue number of the listed items if required after each item. Under the rates column the main contractor is to insert the price for fix only or supply and fix as appropriate.

11 (As 5 above.)

General Note: This Schedule can be used for fix only in which case the supply costs will be covered by a Prime Cost Sum in the summary or for supply and fix by making the appropriate deletions.

27 (As 5 above.)

32 (As 5 above.)

44 (As 5 above.) Insert the closer type.

45 (As 5 above.) Insert the closer type.

46 Insert the window numbers.

47 (As 5 above.) Insert the closer type.

56 The main contractor is to insert the cost of each door and window and the total cost of all fixing or supply and fixing.

Job:	Clause:
Kitchen Unit Schedule	**Rev:**

NOMINATION: Units to be ordered by the Main Contractor from:

RANGE: (Quotation Ref. No):

Code No	Size	Type	Handing	FLAT NO.	FLAT NO.	FLAT NO.	FLAT NO.	FLAT NO.
		Sink Base						
		Floor Unit						
		Worktop						
		Wall Unit						

P.C. SUM FOR SUPPLY INCLUDED IN SUMMARY SHEET.
Main Contractor to price percentage for Profit & Overheads in dealing with order.
PRICES FOR FIXING ARE COVERED IN CLAUSES IN PART III.

1 Insert the job name and reference number. Insert the clause number to follow on from previous schedule.

2 Insert the revision number if necessary.

3 Insert kitchen unit supplier and/or manufacturer.

4 Insert the key numbers for each flat/unit as necessary in right hand box and quotation reference number in left hand box.

6 Complete the schedule for all sink unit specification details.
Insert the numbers of each sink base unit in the right hand box.

13 Complete the schedule for all floor base unit specification details.
Insert the numbers of each base unit in the right hand box.

24 Complete the schedule for all worktop unit specification details.
Insert the numbers of worktops in the right hand box.

33 Complete the schedule for all work unit specification details.
Insert the numbers of wall units in the right hand box.

43 Insert the prime cost sum for units in the summary sheet with other prime costs.

Job:	Clause:	1
Sanitary Fittings Schedule	**Rev:**	2

NOMINATION: Units to be ordered by the Main Contractor from: 3

RANGE: (Quotation Ref. No.):	manufacturer/cat. no.	FLAT	FLAT	FLAT	FLAT	FLAT	4
1 Bath C. iron 1524							5
2 Bath C. iron 1690							
3							
4 Bath Pressed Steel 1524							
5 Bath Pressed Steel 1690							
6							
7 Taps 22mm standard upright							
8 Taps 22mm inclined							
9 Waste 38mm waste top/plug/chain							
10 Waste/combined waste – overflow							
11 Bath panel 1524/1690 black/white							15
12 Bath panel 700 black/white							16
13 Bath panel polished angle strip							
14 Basin 560 × 405							
15 Pedestal/wall hangers/supports							19
16 Basin 510 × 406							
17 Pedestal/wall hangers/supports							21
18 Basin 355 × 255							
19 Basin 500 × 300							
20 Pedestal/wall hangers/supports							24
21 Taps 12mm standard upright							
22 Taps 12mm inclined							
23 WC 'P' trap							
24 WC 'S' trap							
25 Cistern vitreous china							
26 Cistern plastic							
27 WC seat BLACK/WHITE							31
28 Bathroom cabinet							
29 Towel rail (chrome plate)							
30 Toilet roll holder							
31 Mirror 600 × 450							
32 Kitchen sink: 1000 × 500 Drainer LH							
33 1000 × 500 Drainer RH							
34 1000 × 600 Drainer LH							
35 1000 × 600 Drainer RH							
36 1200 × 600 Drainer LH							
37 1200 × 600 Drainer RH							
38 Taps 12mm sink pillar							
39							42
40							
41							

PC SUM FOR SUPPLY INCLUDED IN SUMMARY SHEET
Main Contractor to price percentage for Profit & Overheads in connection with order.
PRICES FOR FIXING ARE COVERED ON CLAUSES IN PART III

46

1 Insert the job name and reference number. Insert the clause number to follow on from previous schedule.
2 Insert the revision number if necessary.

3 Insert the manufacturer's details.

4 Insert the flat key numbers in right hand box.
5 Insert the catalogue numbers for each item of sanitary ware and place the number of items in the right hand box for each flat/unit.

15 Delete black or white as required.
16 " " " " " "

19 Delete pedestal/wall hangers/supports as required.

21 " " " " " " "

24 " " " " " " "

31 Delete black/white as required.

42 Add in additional non standard items as required.

46 Insert the prime sum for fitting in summary sheet with other prime costs.

Appendix

Appendix

The following information has been referred to in the Standard Clauses and Schedules and should be incorporated into the 'Materials and Workmanship' clauses which will be written by the user of this book.

They have been included here to clarify abbreviations and notes used in the text of the book.

Appendix 1 Plaster/Render Mixes
2 Concrete: Nominal Mixes
3 Standard Lintels
4 Definitions of repair abbreviations contained in Window & Door Schedules.

1 Plaster/Render Mixes

Make plaster and render mixes in the proportion given in the table below and use each mix only on the backgrounds listed.

Mix	Background	Undercoat(s) & Thickness	Finish Coat & Thickness	Finish
PLASTER A	Brick adjacent injected DPCs & as directed elsewhere	Protim Premix No. 5–2 equal coats minimum 20 mm	Sirapite 3 mm	Smooth
PLASTER B NEW WORK	New Block/Brick Expanded metal	1:2:9 cement/lime/sand. 1 coat 12 mm pricking up coat and float coat	Sirapite 3 mm	Smooth
PLASTER B REPAIRS	Block/brick/expanded metal lath in repairs, etc	1:2:9 cement/lime/sand. 2 coats to match existing	Sirapite 3 mm	Smooth
PLASTER C	Plasterboard generally	—	Thistleboard Finish 5 mm	Smooth
PLASTER D	Timber lath/dubbing out to plasterboard in small areas	Carlite Bonding	Carlite Finish 2 mm	Smooth
PLASTER E	Existing spalled plaster	Bonding agent	Thistleboard Finish 5 mm	Smooth
RENDER 1	Brick/Block	1:1:6 lime/cement/sand 2 coats of 12 mm or as existing	—	Wood floated textured
RENDER 2	Brick/Block	1:1:6 2 coats of 12 mm or as existing	Dry dash aggregate	

2 Concrete: Nominal Mixes

Make the concrete mixes in the proportions in the table below and use only for the purposes described. Mixes shall have the weight of aggregate shown for every 50Kg of cement.

	A		B
MIX TYPE	1:3:6	1:7 all in	1:2:4
CEMENT TYPE	OPC	OPC	OPC
SAND (Kg)	80	—	80
AGGREGATE (Kg)	190	—	155
ALL-IN-AGGREGATE (Kg)	—	245	—
NOMINAL MAXIMUM AGGREGATE SIZE (mm)	38	38	19
SLUMP (mm)	50–100	50–100	25–50
USE FOR:			
Backfilling excavations, blinding, bedding, drains	X	X	
Mass unreinforced foundations	X	X	
Reinforced foundations/underpinning			X

	A	B
Reinforced and unreinforced site slabs		X
Reinforced lintels, beams, padstones, copings, cills, etc.		X

(Mixes shown are 'Medium workability' to cube strength at 28 days of 21 N/mm²: to CP 114:1969)

3 Standard Lintels

SPAN 0–1 m

Lintel Size	Reinforcement
115 × 150 mm	1 × 12 mm high tensile bar top and bottom, hooked each end. 6mm ms links @ 100 mm c/s
230 × 150 mm	2 × 12 mm high tensile bars (ditto)
280 × 150 mm	3 × 12 mm high tensile bars (ditto)
340 × 150 mm	3 × 12 mm high tensile bars (ditto)

SPAN 1–2 m

Lintel Size	Reinforcement
115 × 230 mm	1 × 16 mm high tensile bars top and bottom, hooked each end. 6 mm ms links @ 150 mm c/s
230 × 230 mm	2 × 16 mm high tensile bars (ditto)
280 × 230 mm	3 × 16 mm high tensile bars (ditto)
340 × 320 mm	3 × 16 mm high tensile bars (ditto)

4 Definitions of Repair Abbreviations used in Window & Door Schedules

As defined in Part III and Schedules for internal doors, external doors and windows, the following definitions apply:

1. Remove ironmongery and make good:
   Cut out and patch in softwood and glue in as required to old or new mortices, etc. and fill ready for decoration.
2. Remove existing door and make good frame and lining: As Definition 1.
3. Repair door by piecing in:
   Cut out defective mouldings and old lock holes, etc. and glue and pin or screw whole patching pieces in sw: fill ready for decoration.
4. Rehang selected door, make good.
   Patch in sw as above (and rehang with new ironmongery).
5. Repair frame/lining by piecing in:
   Cut out and patch in sw defective sections glued and pinned as required and filled to existing profiles ready for decoration.
6. Repair architrave. As Definition 5.
7. New hardwood cill housed to frame:
   Take out cill board and box frame and remove completely old window cill, cut and fit new hardwood cill housed to bottom of box frame to be *full width* of the frame *not* cut between styles. Refix box frame and cill board.
8. Ease and adjust upper/lower sash:
   Remove staff beads, remove sash from cords, shave off/sand down edges as necessary. Renew or refix cords and staff beads as scheduled. Replace rats tails where required. Renew or refix weights as necessary.
9. New sash window to BS644 Part 2 with h.w. cill t.m.e. Cuprinol: Prime, fix:
   The window is to be constructed generally in accordance with the standards of BS644 Part 2 with the following additional requirements:

   (a) Back linings, and rat tails or parting slips to be made from WBP bonded plywood.
   (b) The window to be supplied complete with cast iron pulley wheels, sash cords and weights.
   (c) The hardwood cill to be the full width of the box frame as defined in Definition 7 above.
10. Operations requiring removal/adjustment of sashes and reglazing are to include for the removal and adjustment of sash weights as necessary.